PURE MATHEMATICS

4. DIFFERENTIAL CALCULUS AND APPLICATIONS

Third Edition
By

Anthony Nicolaides

P.A.S.S. PUBLICATIONS

Private Academic & Scientific Studies Limited

© A. NICOLAIDES 1991, 1995, 2007

First Published in Great Britain 1991 by
Private Academic & Scientific Studies Limited

ISBN–13   978–1–872684–97–0   £9–95

THIRD EDITION 2008

This book is copyright under the Berne Convention.
All rights are reserved. Apart as permitted under the Copyright Act, 1956, no part of this publication may be reproduced, stored in a retrieval system, or transmitted in any form of by any means, electronic, electrical, mechanical, optical, photocopying, recording or otherwise, without the prior permission of the publisher.

Titles by the same author.

Revised and Enhanced

1. Algebra.   GCE A Level   ISBN–13   978–1–872684–82–6   £11–95

2. Trigonometry.   GCE A Level   ISBN–13   978–1–872684–87–1   £11–95

3. Complex Numbers.   GCE A Level   ISBN–13   978–1–872684–92–5   £9–95

4. Differential Calculus and Applications.   GCE A Level   ISBN–13   978–1–872684–97–0   £9–95

5. Cartesian and Polar Curve Sketching.   GCE A Level   ISBN–13   978–1–872684–63–5   £9–95

6. Coordinate Geometry in two Dimensions.   GCE A Level   ISBN–13   978–1–872684–68–0   £9–95

7. Integral Calculus and Applications.   GCE A Level   ISBN–13   978–1–872684–73–4   £14–95

8. Vectors in two and three dimensions.   GCE A Level   ISBN–13   978–1–872684–15–4   £9–95

9. Determinants and Matrices.   GCE A Level   ISBN–13   978–1–872684–16–1   £9–95

10. Probabilities.   GCE A Level   ISBN–13   978–1–872684–17–8   £8–95
    This book includes the full solutions.

11. Success in Pure Mathematics: The complete works of GCE A Level. (1–9 above inclusive)   ISBN   978–1–872684–93–2   £39–95

12. Electrical & Electronic Principles. First year Degree Level   ISBN–13   978–1–872684–98–7   £16–95

13. GCSE Mathematics Higher Tier Third Edition.   ISBN–13   978–1–872684–69–7   £19–95

All the books have answers and a CD is attached with FULL SOLUTIONS of all the exercises set at the end of the book.

Preface

This book, which is part of the GCE A level series in Pure Mathematics covers the specialized topic of Differential Calculus and Applications.

The GCE A level series success in Pure Mathematics is comprised of nine books, covering the syllabuses of most examining boards. The books are designed to assist the student wishing to master the subject of Pure Mathematics. The series is easy to follow with minimum help. It can be easily adopted by a student who wishes to study it in the comforts of his home at his pace without having to attend classes formally; it is ideal for the working person who wishes to enhance his knowledge and qualification. The Differential Calculus book, like all the books in the series, the theory is comprehensively dealt with, together with many worked examples and exercises. A step by step approach is adopted in all the worked examples. A CD is attached to the book with FULL SOLUTIONS of all the exercises set at the end of each chapter.

This book develops the basic concepts and skills that are essential for the GCE A level in Pure Mathematics.

Chapter 12 indicates in order the modules of C1, C2, C3, C4, FP1, FP2, FP3 dealing with Differential Calculus and Applications with Questions and Full solutions.

A. Nicolaides

4. DIFFERENTIAL CALCULUS AND APPLICATIONS

CONTENTS

DIFFERENTIATION

1. **ALGEBRAIC FUNCTIONS** 1
    The idea of a limit and the derivative defined as a limit
    The gradient of a tangent as the limit of the gradient of a chord 1
    Concept of a limit 1
    Notation of a gradient. Differentiation from first principles 2
    The derivative of a sum or difference of a function 4
    The derivative of a product of function product rule 5
    The derivative of a quotient of a function. The Quotient rule 7
    The function of a function 8
    Implicit functions 9
    Exercises 1 10

2. **TRIGONOMETRIC OR CIRCULAR FUNCTIONS** 13
    The derivative of $\sin x$ from first principles 13
    The derivative of $\cos x$ from first principles 13
    The derivative of $\tan x$ 13
    The derivative of $\cot x$ 13
    The derivative of $\csc x$ 14
    The derivative of $\sec x$ 14
    Implicit functions 17
    Function of a function of a function 17
    Inverse trigonometric functions 17
    Exercises 2 18

3. **EXPONENTIAL FUNCTIONS** 20
    $y = a^x$, $y = 2^x$, $y = e^x$ 20
    The derivative of e^x from first principles 21
    The derivative of $y = e^{kx}$ 21
    Exercises 3 22

4. **LOGARITHMIC FUNCTIONS** 24
    To determine the derivative of $y = a^x$ 24
    The derivative of $y = \ln kx$ 25
    Exercises 4 25

5. **HYPERBOLIC FUNCTION** 27
    Function of a function 28
    Inverse hyperbolic functions 29
    Exercises 5 31

6. **PARAMETRIC EQUATIONS** 33
    Exercises 6

APPLICATIONS OF DIFFERENTIATION — 37

7. SECOND AND HIGHER DERIVATIVES OF A FUNCTION — 37
  Notation of second derivative — 37
  The meaning of the second derivative — 37
  Exercises 7

8. TANGENTS AND NORMALS — 40
  To find the angle between two lines — 40
  Exercises 8 — 44

9. SMALL INCREMENTS AND APPROXIMATION RATES — 45
  L'HÔPITAL'S Rule — 46
  Small increments — 46
  Rate of change — 48
  Exercises 9 — 49

10. NEWTON-RAPHSON APPROXIMATION — 51
  Methods of approximation to the solution of an equation, improvements of the value of such approximation including the Newton-Raphson method — 51
  Exercises 10 — 54

11. MACLAURIN'S EXPANSIONS — 55
  Power series. — 55
  Third and higher order derivatives — 56
  Quadratic and higher degree polynomial approximations to simple functions — 56
  Taylor's Theorem — 57
  Approximate solution of equations — 57
  Successive approximations — 60
  Numerical methods for the solution of Differential Equations Polynomial Approximations using Taylor series — 63
  Exercises 11 — 67

12. ADDITIONAL EXAMPLES FOR C1, C2, C3, C4, FP1, FP2, FP3 — 69

  MISCELLANEOUS — 78

  ANSWERS — 85

  INDEX — 96

Algebraic Functions

The idea of a limit and the derivative defined as a limit. The gradient of a tangent as the limit of the gradient of a chord.

Consider the general form

$$y = ax^n$$

where a and n are constants and x and y are variables, x is the independent variable and y is the dependent variable.

If $a = 1$ and $n = 2$ then

$$y = x^2$$

the graph takes the form of a parabola as shown in Fig. 4-I/1 for $x \geq 0$.

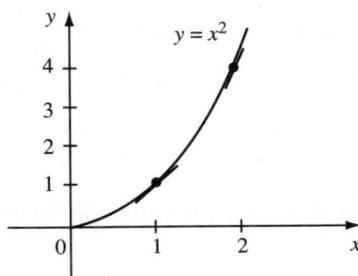

Fig. 4-I/1

It is required to find the gradient at a point, say, $x = 2$. A tangent is drawn at $x = 2$ as shown, the gradient is defined as the ratio of the vertical distance, 4, to the horizontal distance, 1, that is, gradient $= \dfrac{4}{1} = 4$.

The gradient at $x = 1$, is again found as $\dfrac{2}{1} = 2$.

The method of finding gradients at different points of a curve is rather approximate and tedious. A much neater method is used in finding gradients, the <u>differential calculus</u> method.

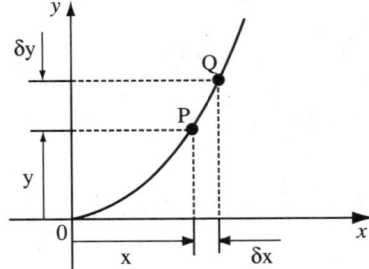

Fig. 4-I/2

Consider a point P on the parabola, as shown in Fig. 4-I/2 of coordinates $P(x, y)$. A point Q close to P has coordinates $Q(x + \delta x, y + \delta y)$ where δx and δy are infinitesimally small distances along the x and y axes respectively.

Expanding Fig. 4-I/2 for convenience and clear illustration as in Fig. 4-I/3, the chord PQ is drawn. The gradient

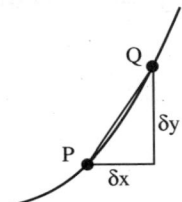

Fig. 4-I/3

of the chord **PQ** is given as $\left(\dfrac{\delta y}{\delta x}\right)$, that is, gradient of

$$\mathbf{PQ} = \dfrac{\delta y}{\delta x}.$$

Concept of a Limit

Consider a length of 1 metre, halving this length gives a length of 500 millimetres, halving this again gives a

1

length of 250 mm, halving this gives a length of 125 mm, keep on halving the lengths indefinitely the length will approach a very small length such as $\delta x \to 0$ (delta x tends to zero).

If we apply this concept to Fig. 4-I/3 when **Q** approaches **P** and $\delta x \to 0$ then the gradient of the chord **PQ** will approach the gradient of the tangent at **P**, Fig. 4-I/4.

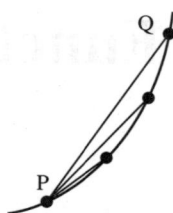

Fig. 4-I/4

Gradient at $\mathbf{P} = \left(\dfrac{\delta y}{\delta x}\right)_{\delta x \to 0}$ gradient at $\mathbf{P} = \dfrac{dy}{dx}$

$\dfrac{dy}{dx}$ is the ratio $\dfrac{\delta y}{\delta x}$ as δx tends to zero.

Notation of the Gradient

The gradient of a function is denoted as $\dfrac{dy}{dx}$.

This has alternative terminology, the <u>gradient of the tangent at **P**</u>, the <u>slope</u> of the tangent at **P**, <u>the first derivative</u>, or the <u>primitive</u> of the function.

To Derive the First Derivative of the Function $y = x^2$ from First Principles

If x is increased to $x + \delta x$, then the corresponding value of y is $y + \delta y$.

$$y + \delta y = (x + \delta x)^2 \qquad \ldots (1)$$
$$y = x^2 \qquad \ldots (2)$$

subtracting (2) from (1)

$$\delta y = (x + \delta x)^2 - x^2$$
$$\delta y = x^2 + 2x\,\delta x + \delta x^2 - x^2$$
$$\delta y = 2x\,\delta x + \delta x^2$$

dividing each term by δx

$$\dfrac{\delta y}{\delta x} = 2x + \delta x \qquad \ldots (3)$$

applying the idea of the limit, as $\delta x \to 0$ $\dfrac{\delta y}{\delta x} \to \dfrac{dy}{dx}$ in (3), we have

$$\boxed{\dfrac{dy}{dx} = 2x}$$

To Derive the First Derivative of the Function $y = 2x^3$ from First Principles

If x is increased to $x + \delta x$, then the corresponding value of y is $y + \delta y$

$$y + \delta y = 2(x + \delta x)^3 \qquad \ldots (1)$$
$$y = 2x^3 \qquad \ldots (2)$$

substracting (2) from (1)

$$\delta y = 2(x + \delta x)^3 - 2x^3$$
$$\delta y = 2(x^3 + 3x^2\,\delta x + 3x\delta x^2 + \delta x^3) - 2x^3$$
$$\delta y = 2x^3 + 6x^2\,\delta x + 6x\delta x^2 + 2\delta x^3 - 2x^3$$
$$\delta y = 6x^2\,\delta x + 6x\delta x^2 + 2\delta x^3$$

dividing each term by δx, we have

$$\dfrac{\delta y}{\delta x} = 6x^2 + 6x\delta x + 2\delta x^2 \qquad \ldots (3)$$

applying the idea of the limit, as $\delta x \to 0$, $\dfrac{\delta y}{\delta x} \to \dfrac{dy}{dx}$ in (3), we have

$$\boxed{\dfrac{dy}{dx} = 6x^2}$$

From the last derivations, it can be deduced that the derivative of $y = ax^n$ is

$$\boxed{\dfrac{dy}{dx} = anx^{n-1}}$$

Observe that the coefficient of x^n is multiplied by the power n and the power is reduced by unity.

$$y = x^2, \dfrac{dy}{dx} = 1 \times 2x^{2-1} = 2x$$
$$y = 2x^3, \dfrac{dy}{dx} = 2 \times 3x^{3-1} = 6x^2$$
$$y = ax^n, \dfrac{dy}{dx} = anx^{n-1}.$$

Algebraic Functions — 3

WORKED EXAMPLE 1

Derive from first principles the derivatives of the functions:-

(i) $y = 3x$
(ii) $y = 2x^2$
(iii) $y = 5x^3$
(iv) $y = \dfrac{5}{x}$
(v) $y = 3x^{\frac{1}{2}}$.

Solution 1

(i) $y = 3x$ ...(1)

$y + \delta y = 3(x + \delta x)$ ...(2)

(2) − (1)

$\delta y = 3(x + \delta x) - 3x$

$\delta y = 3x + 3\delta x - 3x$

$\delta y = 3\delta x$

dividing each side by δx

$\dfrac{\delta y}{\delta x} = 3$

as $\delta x \to 0$, $\dfrac{\delta y}{\delta x} \to \dfrac{dy}{dx}$

$\boxed{\dfrac{dy}{dx} = 3}$

(ii) $y = 2x^2$ ...(1)

$y + \delta y = 2(x + \delta x)^2$ ...(2)

(2) − (1)

$\delta y = 2x^2 + 4x\delta x + 2\delta x^2 - 2x^2$

$\delta y = 4x\delta x + 2\delta x^2$

dividing each term by δx, we have

$\dfrac{\delta y}{\delta x} = 4x + 2\delta x$

as $\delta x \to 0$, $\dfrac{\delta y}{\delta x} \to \dfrac{dy}{dx}$

$\boxed{\dfrac{dy}{dx} = 4x}$

(iii) $y = 5x^3$ ...(1)

$y + \delta y = 5(x + \delta x)^3$ ...(2)

$\delta y = 5(x + \delta x)^3 - 5x^3$

$= 5\left(x^3 + 3x^2\delta x + 3x\delta x^2 + \delta x^3\right) - 5x^3$

$\delta y = 5x^3 + 15x^2\delta x + 15x\delta x^2 + 5\delta x^3 - 5x^3$

$\delta y = 15x^2\delta x + 15x\delta x^2 + 5\delta x^3$

dividing each term by δx

$\dfrac{\delta y}{\delta x} = 15x^2 + 15x\delta x + 5\delta x^2$ as $\delta x \to 0$,

$\dfrac{\delta y}{\delta x} \to \dfrac{dy}{dx}$ $\dfrac{dy}{dx} = 15x^2$

(iv) $y = \dfrac{5}{x}$ ...(1)

$y + \delta y = \dfrac{5}{x + \delta x}$ ...(2)

$\delta y = \dfrac{5}{x + \delta x} - \dfrac{5}{x}$

$\delta y = \dfrac{5x - 5(x + \delta x)}{x(x + \delta x)} = \dfrac{5x - 5x - 5\delta x}{x(x + \delta x)}$

$\delta y = \dfrac{-5\delta x}{x(x + \delta x)}$

dividing each term by δx $\dfrac{\delta y}{\delta x} = \dfrac{-5}{x(x + \delta x)}$ as

$\delta x \to 0$, $\dfrac{\delta y}{\delta x} \to \dfrac{dy}{dx}$ $\dfrac{dy}{dx} = -\dfrac{5}{x^2}$.

Alternatively, $y = \dfrac{5}{x} = 5x^{-1}$ ...(1)

$y + \delta y = 5(x + \delta x)^{-1}$ ...(2)

(2) − (1) $\delta y = 5(x + \delta x)^{-1} - 5x^{-1}$; using the binomial expansion

$(x + \delta x)^{-1} = x^{-1}\left(1 + \dfrac{\delta x}{x}\right)^{-1}$

$= x^{-1}\left(1 + (-1)\dfrac{\delta x}{x} + (-1)(-2)\left(\dfrac{\delta x}{x}\right)^2 \dfrac{1}{2!} + \ldots\right)$

$= x^{-1}\left(1 - \dfrac{\delta x}{x} + \dfrac{\delta x^2}{x^2} + \ldots\right)$

$$\delta y = 5x^{-1}\left(1 - \frac{\delta x}{x} + \frac{\delta x^2}{x^2} + \ldots\right) - 5x^{-1}$$

$$= -5x^{-1}\frac{\delta x}{x} + 5x^{-1}\frac{\delta x^2}{x^2} + \ldots$$

dividing each term by δx

$$\frac{\delta y}{\delta x} = -\frac{5x^{-1}}{x} + 5x^{-1}\frac{\delta x}{x^2} + \ldots \text{ as } \delta x \to 0,$$

$$\frac{\delta y}{\delta x} \to \frac{dy}{dx} \qquad \frac{dy}{dx} = \frac{-5}{x^2}.$$

(v) $y = 3x^{\frac{1}{2}}$ ...(1)

$y + \delta y = 3(x + \delta x)^{\frac{1}{2}}$ ...(2)

(2) − (1) $\delta y = 3(x + \delta x)^{\frac{1}{2}} - 3x^{\frac{1}{2}}$

using the binomial expansion

$(x + \delta x)^{\frac{1}{2}}$

$$= x^{\frac{1}{2}}\left(1 + \frac{\delta x}{x}\right)^{\frac{1}{2}}$$

$$= x^{\frac{1}{2}}\left(1 + \frac{1}{2}\frac{\delta x}{x} + \frac{1}{2}\left(-\frac{1}{2}\right)\left(\frac{\delta x}{x}\right)^2\frac{1}{2!} + \ldots\right)$$

$$= x^{\frac{1}{2}} + \frac{1}{2}x^{\frac{1}{2}}\frac{\delta x}{x} - \frac{1}{8}\frac{(\delta x)^2}{x^2}x^{\frac{1}{2}} + \ldots$$

$$\delta y = 3\left(x^{\frac{1}{2}} + \frac{1}{2}x^{\frac{1}{2}}\frac{\delta x}{x}\right.$$
$$\left. - \frac{1}{8}\left(\frac{\delta x}{x}\right)^2 x^{\frac{1}{2}} + \ldots\right) - 3x^{\frac{1}{2}}$$

$$= \frac{3}{2}x^{\frac{1}{2}}\frac{\delta x}{x} - \frac{3}{8}\left(\frac{\delta x}{x}\right)^2 x^{\frac{1}{2}} + \ldots$$

dividing each term by δx

$$\frac{\delta y}{\delta x} = \frac{3}{2}\frac{x^{\frac{1}{2}}}{x} - \frac{3}{8}\frac{\delta x}{x^2}x^{\frac{1}{2}} + \ldots \text{ as } \delta x \to 0,$$

$$\frac{\delta y}{\delta x} \to \frac{dy}{dx} \qquad \frac{dy}{dx} = \frac{3}{2}\frac{x^{\frac{1}{2}}}{x} = \frac{3}{2x^{\frac{1}{2}}}.$$

WORKED EXAMPLE 2

Determine the derivatives of the functions:-

(i) $y = 3x$

(ii) $y = 2x^2$

(iii) $y = 5x^3$

(iv) $y = \dfrac{5}{x}$

(v) $y = 3x^{\frac{1}{2}}$, using the formula

$$\frac{dy}{dx} = anx^{n-1}, \text{ when } y = ax^n.$$

Solution 2

(i) $y = 3x^1 \qquad \dfrac{dy}{dx} = 3 \times 1x^{1-1} = 3x^0 = 3$

(ii) $y = 2x^2 \qquad \dfrac{dy}{dx} = 2 \times 2x^{2-1} = 4x$

(iii) $y = 5x^3 \qquad \dfrac{dy}{dx} = 5 \times 3x^{3-1} = 15x^2$

(iv) $y = \dfrac{5}{x} = 5x^{-1}$ expressing in the form $y = ax^n$

$$\frac{dy}{dx} = 5(-1)x^{-1-1} = -5x^{-2} = -\frac{5}{x^2}$$

(v) $y = 3x^{\frac{1}{2}}$

$$\frac{dy}{dx} = 3\left(\frac{1}{2}\right)x^{\frac{1}{2}-1} = \frac{3}{2}x^{-\frac{1}{2}} = \frac{3}{2x^{\frac{1}{2}}}.$$

To Determine the Derivative of a Sum or Difference of a Function

Let $y = f(x) + g(x) - p(x) - q(x)$ ...(1)

where f(x) denotes a function of x and g(x), p(x), and q(x) are different functions of x. The derivative of the function (1) is

$$\frac{dy}{dx} = \frac{d}{dx}f(x) + \frac{d}{dx}g(x) - \frac{d}{dx}p(x) - \frac{d}{dx}q(x) \text{ or simply}$$

$$\frac{dy}{dx} = f'(x) + g'(x) - p'(x) - q'(x)$$

where f'(x) denotes the first derivative of f(x) and similarly g'(x), p'(x) and q'(x) denote the first derivatives of the functions $g(x)$, $p(x)$ and $q(x)$ respectively. The differentiation across addition and subtraction is <u>distributive</u>.

WORKED EXAMPLE 3

Determine the derivatives of the functions:
(i) $y = 3x - 3x^2 - 5x^3$
(ii) $y = \dfrac{2}{x} - 3x + 4x^2$.

Solution 3

(i) $y = 3x - 3x^2 - 5x^3$

$$\dfrac{dy}{dx} = \dfrac{d}{dx}(3x) - \dfrac{d}{dx}(3x^2) - \dfrac{d}{dx}(5x^3)$$

$$= 3 - 6x - 15x^2$$

(ii) $y = \dfrac{2}{x} - 3x + 4x^2 = 2x^{-1} - 3x + 4x^2$

$$\dfrac{dy}{dx} = -2x^{-2} - 3 + 8x = -\dfrac{2}{x^2} - 3 + 8x.$$

WORKED EXAMPLE 4

Determine the gradients of the following functions at the points adjacent to each function

(i) $y = -2x^{\frac{1}{2}}$ $(x = 1)$
(ii) $y = 3x^{\frac{1}{3}}$ $(x = -1)$
(iii) $y = \dfrac{1}{\sqrt{x}}$ $\left(x = \dfrac{1}{2}\right)$
(iv) $y = \sqrt[3]{x^2}$ $(x = 1)$
(v) $y = x^2 - x^3 + x^4$ $(x = 0)$
(vi) $y = \dfrac{1}{x} - \dfrac{1}{x^2} + \dfrac{1}{x^3}$ $(x = 1)$.

Solution 4

(i) $y = -2x^{\frac{1}{2}}$ $\dfrac{dy}{dx} = -x^{-\frac{1}{2}} = -\dfrac{1}{x^{\frac{1}{2}}}$

at $x = 1$ $\dfrac{dy}{dx} = -\dfrac{1}{1} = -1$

(ii) $y = 3x^{\frac{1}{3}}$ $\dfrac{dy}{dx} = x^{-\frac{2}{3}} = \dfrac{1}{x^{\frac{2}{3}}}$

at $x = -1$ $\dfrac{dy}{dx} = \dfrac{1}{\sqrt[3]{x^2}} = \dfrac{1}{\sqrt[3]{(-1)^2}} = 1$

(iii) $y = \dfrac{1}{\sqrt{x}} = x^{-\frac{1}{2}}$

$\dfrac{dy}{dx} = -\dfrac{1}{2}x^{-\frac{3}{2}} = -\dfrac{1}{2\sqrt{x^3}}$ at $x = \dfrac{1}{2}$

$\dfrac{dy}{dx} = -\dfrac{1}{2\sqrt{\dfrac{1}{8}}} = -\dfrac{1}{\sqrt{\dfrac{4}{8}}} = -\sqrt{2}.$

(iv) $y = \sqrt[3]{x^2} = x^{\frac{2}{3}}$ $\dfrac{dy}{dx} = \dfrac{2}{3}x^{-\frac{1}{3}}$

when $x = 1$, $\dfrac{dy}{dx} = \dfrac{2}{3}$

(v) $y = x^2 - x^3 + x^4$ $\dfrac{dy}{dx} = 2x - 3x^2 + 4x^3$

when $x = 0$, $\dfrac{dy}{dx} = 0.$

(vi) $y = \dfrac{1}{x} - \dfrac{1}{x^2} + \dfrac{1}{x^3} = x^{-1} - x^{-2} + x^{-3}$

$\dfrac{dy}{dx} = -x^{-2} + 2x^{-3} - 3x^{-4}$

$= -\dfrac{1}{x^2} + \dfrac{2}{x^3} - \dfrac{3}{x^4}$

when $x = 1$, $\dfrac{dy}{dx} = -\dfrac{1}{1} + \dfrac{2}{1} - \dfrac{3}{1} = -2.$

The Derivative of a Product of a Function Product Rule

Let $u = f(x)$ and $v = g(x)$

$$y = u \cdot v$$

the derivative of a product

$$\boxed{\dfrac{dy}{dx} = v\dfrac{du}{dx} + u\dfrac{d}{dx}v}$$

WORKED EXAMPLE 5

Determine the derivatives of the following products using the product rule.

(i) $y = x(x^2 + 1)$

(ii) $y = x^{\frac{1}{2}}(\sqrt{x} + 1)$

(iii) $y = (x+1)(x^3 - 1)$

(iv) $y = (3x + 5)(5x^2 - 7)$.

Solution 5

(i) $y = x(x^2 + 1)$ where $u = x$, $v = x^2 + 1$

$$\frac{du}{dx} = 1, \quad \frac{dv}{dx} = 2x$$

$$\frac{dy}{dx} = (x^2 + 1)1 + x2x = x^2 + 1 + 2x^2 = 3x^2 + 1$$

(ii) $y = x^{\frac{1}{2}}(\sqrt{x} + 1)$ where $u = x^{\frac{1}{2}}$

$$\frac{du}{dx} = \frac{1}{2}x^{-\frac{1}{2}}$$

$$v = \sqrt{x} + 1 = x^{\frac{1}{2}} + 1$$

$$\frac{dv}{dx} = \frac{1}{2}x^{-\frac{1}{2}}$$

$$\frac{dy}{dx} = \left(x^{\frac{1}{2}} + 1\right)\frac{1}{2}x^{-\frac{1}{2}} + x^{\frac{1}{2}}\frac{1}{2}x^{-\frac{1}{2}}$$

$$= \frac{1}{2} + \frac{1}{2}x^{-\frac{1}{2}} + \frac{1}{2} = 1 + \frac{1}{2}x^{-\frac{1}{2}}$$

$$= 1 + \frac{1}{2x^{\frac{1}{2}}}$$

(iii) $y = (x+1) \cdot (x^3 - 1)$ where $u = x + 1$,

$$\frac{du}{dx} = 1 \quad v = x^3 - 1 \quad \frac{dv}{dx} = 3x^2$$

$$\frac{dy}{dx} = \left(x^3 - 1\right)1 + (x+1)3x^2$$

$$= x^3 - 1 + 3x^3 + 3x^2 = 4x^3 + 3x^2 - 1$$

(iv) $y = (3x + 5)(5x^2 - 7)$ where $u = 3x + 5$,

$$\frac{du}{dx} = 3 \text{ and } \quad v = 5x^2 - 7, \quad \frac{dv}{dx} = 10x$$

$$\frac{dy}{dx} = \left(5x^2 - 7\right)3 + (3x+5)10x$$

$$= 15x^2 - 21 + 30x^2 + 50x$$

$$= 45x^2 + 50x - 21.$$

WORKED EXAMPLE 6

Determine the derivatives of the following functions, using the rule $y = ax^n$, $\frac{dy}{dx} = anx^{n-1}$.

(i) $y = x(x^2 + 1)$

(ii) $y = x^{\frac{1}{2}}(\sqrt{x} + 1)$

(iii) $y = (x+1)(x^3 - 1)$

(iv) $y = (3x + 5)(5x^2 - 7)$.

Solution 6

(i) $y = x(x^2 + 1) = x^3 + x$,

$$\frac{dy}{dx} = 3x^2 + 1$$

(ii) $y = x^{\frac{1}{2}}\left(x^{\frac{1}{2}} + 1\right) = x + x^{\frac{1}{2}}$,

$$\frac{dy}{dx} = 1 + \frac{1}{2}x^{-\frac{1}{2}} = 1 + \frac{1}{2x^{\frac{1}{2}}}$$

(iii) $y = (x+1)(x^3 - 1) = x^4 + x^3 - x - 1$

$$\frac{dy}{dx} = 4x^3 + 3x^2 - 1$$

(iv) $y = (3x+5)(5x^2 - 7) = 15x^3 + 25x^2 - 21x - 35$

$$\frac{dy}{dx} = 45x^2 + 50x - 21$$

Observe that the results are the same as in example 5.

Algebraic Functions

The Derivative of a Quotient of a Function The Quotient Rule

$$y = \frac{u}{v}$$

where $u = f(x)$ and $v = g(x)$ the quotient rule is given by

$$\boxed{\frac{dy}{dx} = \frac{v\dfrac{du}{dx} - u\dfrac{dv}{dx}}{v^2}}$$

Worked Example 7

Determine the derivatives of the quotients

(i) $y = \dfrac{x}{x^2 + 1}$

(ii) $y = \dfrac{3x - 1}{5x^2 - 3}$.

Solution 7

(i) $y = \dfrac{x}{x^2 + 1}$ where $u = x$

and $v = x^2 + 1$, $\quad \dfrac{du}{dx} = 1 \quad$ and $\quad \dfrac{dv}{dx} = 2x$

$$\frac{dy}{dx} = \frac{(x^2 + 1) \cdot 1 - x \cdot 2x}{(x^2 + 1)^2}$$

$$= \frac{x^2 + 1 - 2x^2}{(x^2 + 1)^2} = \frac{1 - x^2}{(x^2 + 1)^2}$$

the result should be always simplified.

(ii) $y = \dfrac{3x - 1}{5x^2 - 3}$ where $u = 3x - 1$ and $v = 5x^2 - 3$,

$\dfrac{du}{dx} = 3$ and $\dfrac{dv}{dx} = 10x$

$$\frac{dy}{dx} = \frac{(5x^2 - 3) \cdot 3 - (3x - 1) \cdot 10x}{(5x^2 - 3)^2}$$

$$= \frac{15x^2 - 9 - 30x^2 + 10x}{(5x^2 - 3)^2}$$

$$= \frac{-15x^2 + 10x - 9}{(5x^2 - 3)^2}$$

in simplified form.

Worked Example 8

Determine the derivatives of the functions:-

(i) $y = \dfrac{x^2 + 3}{x}$

(ii) $y = \dfrac{x^3 - 5}{\sqrt{x}}$ using

(a) the general rule,

(b) the quotient rule.

Solution 8

(a) (i) $y = \dfrac{x^2 + 3}{x} = x + 3x^{-1}$

$$\frac{dy}{dx} = 1 - 3x^{-2} = 1 - \frac{3}{x^2}$$

(ii) $y = \dfrac{x^3 - 5}{\sqrt{x}} = (x^3 - 5)x^{-\frac{1}{2}}$

$$= x^{\frac{5}{2}} - 5x^{-\frac{1}{2}}$$

$$\frac{dy}{dx} = \frac{5}{2}x^{\frac{3}{2}} + \frac{5}{2}x^{-\frac{3}{2}}.$$

(b) (i) $y = \dfrac{x^2 + 3}{x}$ where $u = x^2 + 3$, $\quad v = x$

$\dfrac{du}{dx} = 2x, \quad \dfrac{dv}{dx} = 1$

$$\frac{dy}{dx} = \frac{x \cdot 2x - (x^2 + 3) \cdot 1}{x^2}$$

$$= \frac{2x^2 - x^2 - 3}{x^2} = \frac{x^2 - 3}{x^2}$$

$$= 1 - \frac{3}{x^2}.$$

(ii) $y = \dfrac{x^3 - 5}{\sqrt{x}}$ where $u = x^3 - 5$, $\dfrac{du}{dx} = 3x^2$

$v = x^{\frac{1}{2}} \qquad \dfrac{dv}{dx} = \dfrac{1}{2}x^{-\frac{1}{2}}$

$$\frac{dy}{dx} = \frac{x^{\frac{1}{2}} \cdot 3x^2 - (x^3 - 5)\dfrac{1}{2}x^{-\frac{1}{2}}}{x}$$

$$= \frac{3x^{\frac{5}{2}} - \frac{1}{2}x^{\frac{5}{2}} + \frac{5}{2}x^{-\frac{1}{2}}}{x}$$

$$= \frac{\frac{5}{2}x^{\frac{5}{2}} + \frac{5}{2}x^{-\frac{1}{2}}}{x}$$

$$= \frac{5}{2}x^{\frac{3}{2}} + \frac{5}{2}x^{-\frac{3}{2}}.$$

The Function of a Function

This is best illustrated by an example.

WORKED EXAMPLE 9

Determine the derivative $y = (3x^2 + 5)^{23}$.

Solution 9

$y = (3x^2 + 5)^{23}$.

Let $u = 3x^2 + 5$, $\dfrac{du}{dx} = 6x$

$y = u^{23}$, $\dfrac{dy}{du} = 23u^{22}$

$\dfrac{dy}{dx} = \dfrac{dy}{du} \cdot \dfrac{du}{dx} = 23u^{22} \cdot 6x$

$= 23(3x^2 + 5)^{22} \, 6x$

$= 138x(3x^2 + 5)^{22}$

this method is called function of a function (sometimes called in bed function), that is, y is a function of u which is a function of x.

WORKED EXAMPLE 10

Determine the derivatives of the functions:-

(i) $y = \dfrac{x^2 - 1}{(x^2 + 1)^{15}}$

(ii) $y = (3x^2 - 1)^{\frac{1}{2}}$

(iii) $y = \sqrt{5x - 3}$.

Solution 10

(i) $y = \dfrac{x^2 - 1}{(x^2 + 1)^{15}}$

let $u = x^2 - 1$ and $v = (x^2 + 1)^{15}$

$\dfrac{du}{dx} = 2x$, $v = (x^2 + 1)^{15} = W^{15}$

where $W = x^2 + 1$

$\dfrac{dW}{dx} = 2x$, $\dfrac{dv}{dW} = 15W^{14} = 15(x^2 + 1)^{14}$

$\dfrac{dv}{dx} = 30x(x^2 + 1)^{14}$

$\dfrac{dy}{dx} = \dfrac{(x^2 + 1)^{15} \cdot 2x - (x^2 - 1) \cdot 30x(x^2 + 1)^{14}}{(x^2 + 1)^{30}}$

$= \dfrac{(x^2 + 1)^{14} \left[(x^2 + 1) 2x - 30x^3 + 30x\right]}{(x^2 + 1)^{30}}$

$= \dfrac{2x^3 + 2x - 30x^3 + 30x}{(x^2 + 1)^{16}} = \dfrac{-28x^3 + 32x}{(x^2 + 1)^{16}}$

$= \dfrac{4x(8 - 7x^2)}{(x^2 + 1)^{16}}.$

(ii) $y = (3x^2 - 1)^{\frac{1}{2}}$

let $u = 3x^2 - 1 \Rightarrow \dfrac{du}{dx} = 6x$

$y = u^{\frac{1}{2}}$, $\dfrac{dy}{du} = \dfrac{1}{2}u^{-\frac{1}{2}}$

$\dfrac{dy}{dx} = \dfrac{dy}{du} \cdot \dfrac{du}{dx} = \dfrac{1}{2}u^{-\frac{1}{2}}(6x) = 3x(3x^2 - 1)^{-\frac{1}{2}}$

(iii) $y = \sqrt{5x - 3}$ let $u = 5x - 3 \Rightarrow \dfrac{du}{dx} = 5$

$y = u^{\frac{1}{2}}$, $\dfrac{dy}{du} = \dfrac{1}{2}u^{-\frac{1}{2}}$

$\dfrac{dy}{dx} = \dfrac{dy}{du} \cdot \dfrac{du}{dx}$

$= \dfrac{1}{2}(5x - 3)^{-\frac{1}{2}}(5) = \dfrac{5}{2\sqrt{5x - 3}}$

Implicit Functions

We have seen that y is expressed in terms of x explicitly, such as $y = \dfrac{1}{x}$. This is expressed also implicitly such as $xy = 1$.

WORKED EXAMPLE 11

(a) Find $\dfrac{dy}{dx}$ for the explicit function $y = \dfrac{1}{x}$.

(b) Find $\dfrac{dy}{dx}$ for the implicit function $xy = 1$.

Solution 11

(a) $y = \dfrac{1}{x} = x^{-1}$ $\qquad \dfrac{dy}{dx} = -x^{-2} = -\dfrac{1}{x^2}$

(b) $xy = 1$

differentiating with respect to x, we have

$$\dfrac{d}{dx}(xy) = \dfrac{d}{dx}(1)$$

$$\dfrac{dx}{dx}y + x\dfrac{dy}{dx} = 0$$

$$y + x\dfrac{dy}{dx} = 0$$

$$x\dfrac{dy}{dx} = -y$$

$$\dfrac{dy}{dx} = -\dfrac{y}{x} = -\dfrac{\frac{1}{x}}{x}$$

$$= -\dfrac{1}{x^2}.$$

The answers are the same as expected, but the second method is more difficult when we use the function implicitly.

WORKED EXAMPLE 12

Determine $\dfrac{dy}{dx}$ and $\dfrac{dx}{dy}$ for the implicit function $\dfrac{x^2}{4} + \dfrac{y^2}{9} = 1$.

Solution 12

$$\dfrac{x^2}{4} + \dfrac{y^2}{9} = 1.$$

Differentiating with respect to x

$$\dfrac{d}{dx}\left(\dfrac{x^2}{4}\right) + \dfrac{d}{dx}\left(\dfrac{y^2}{9}\right) = \dfrac{d}{dx}(1)$$

$$\dfrac{2x}{4}\dfrac{dx}{dx} + \dfrac{2y}{9}\dfrac{dy}{dx} = 0$$

$$\dfrac{x}{2} + \dfrac{2}{9}y\dfrac{dy}{dx} = 0, \qquad \dfrac{dy}{dx} = -\dfrac{\frac{x}{2}}{\left(\frac{2y}{9}\right)} = -\dfrac{9x}{4y}$$

$$\dfrac{dy}{dx} = -\dfrac{9x}{4y}, \text{ this is expressed in terms of } x \text{ and } y$$

$$\dfrac{x^2}{4} + \dfrac{y^2}{9} = 1$$

differentiating with respect to y

$$\dfrac{d}{dy}\left(\dfrac{x^2}{4}\right) + \dfrac{d}{dy}\left(\dfrac{y^2}{9}\right) = \dfrac{d}{dy}(1)$$

$$\dfrac{2x}{4}\dfrac{dx}{dy} + \dfrac{2y}{9}\dfrac{dy}{dy} = 0$$

$$\dfrac{1}{2}x\dfrac{dx}{dy} + \dfrac{2}{9}y = 0, \qquad \dfrac{dx}{dy} = -\dfrac{\frac{2y}{9}}{\frac{x}{2}}$$

$$\dfrac{dx}{dy} = -\dfrac{4y}{9x}, \text{ the reciprocal of this agrees with}$$
$$\dfrac{dy}{dx} = -\dfrac{9x}{4y}.$$

WORKED EXAMPLE 13

Differentiate with respect to x the function $xy + y^2x^2 = 3$. Determine $\dfrac{dx}{dy}$ and check the answer.

Solution 13

$$xy + y^2x^2 = 3 \qquad \ldots (1)$$

differentiating with respect to x, we have

1. $y + x\dfrac{dy}{dx} + 2y\dfrac{dy}{dx}x^2 + y^2 2x = 0$

$\dfrac{dy}{dx}(x + 2x^2 y) = -y^2 2x - y$

$\dfrac{dy}{dx} = -\dfrac{y(1 + 2xy)}{x(1 + 2xy)} = -\dfrac{y}{x}$

differentiating equation (1) with respect to y, we have

$\dfrac{dx}{dy} \cdot y + x \cdot 1 + 2yx^2 + y^2 2x \dfrac{dx}{dy} = 0$

$\dfrac{dx}{dy}(y + 2xy^2) = -(x + 2x^2 y)$

$\dfrac{dx}{dy} = -\dfrac{x(1 + 2xy)}{y(1 + 2xy)} = -\dfrac{x}{y}$

the reciprocal gives $\dfrac{dy}{dx} = -\dfrac{y}{x}$.

Exercises 1

1. Write down the derivative of the general algebraic form $y = ax^n$.

2. Differentiate from first principles the following algebraic functions:-
   (i) $y = 3$
   (ii) $y = x$
   (iii) $y = -x^2 + 1$
   (iv) $y = 2x^3$
   (v) $y = 5x - \dfrac{3}{x} + \dfrac{1}{x^2}$.

3. Differentiate the following algebraic functions:
   (i) $y = 3x$
   (ii) $y = 5$
   (iii) $y = \dfrac{3}{x}$
   (iv) $y = -x^2 - x^3 - x^4$
   (v) $y = \dfrac{3}{\sqrt{x}}$
   (vi) $y = \dfrac{1}{x} + \dfrac{4}{x^2} - \dfrac{3}{x^3}$
   (vii) $x = 3t^2 - 5t$
   (viii) $\Theta = 3t^2 - 5t$
   (ix) $r = \dfrac{1}{t} + t - t^2$
   (x) $Z = 5y^2 - 5y^3 - 7y^5$.

4. Differentiate the following products:-
   (i) $y = (x^2 + 3)(x^2 - 5)$
   (ii) $t = (x + 1)(x^3 - 9)$
   (iii) $\Theta = (3t^3 - 5)t^5$
   (iv) $y = 3(x^{27} - 3)$
   (v) $y = x(x^2 - 1)(x^3 - 2)$.

5. Differentiate the following quotients:-
   (i) $y = \dfrac{x^2}{x^5 - 1}$
   (ii) $Z = \dfrac{t^3 - 1}{t^3 + 1}$
   (iii) $\Theta = \dfrac{3r}{r^4 - 1}$
   (iv) $y = \dfrac{x^4 - 3x^3 + 2x^2}{5x}$
   (v) $y = \dfrac{x - 1}{x + 2}$.

6. Differentiate the following functions:-
   (i) $y = \sqrt{3x + 1}$
   (ii) $y = \dfrac{1}{\sqrt{x - 1}}$
   (iii) $y = (x^3 - 1)^3$
   (iv) $y = (x^2 - 1)^3 (x^3 + 1)^4$
   (v) $y = (5x^2 - 7)^{\frac{1}{3}}$.

7. Distinguish between implicit and explicit functions.

8. Differentiate the following:-
   (i) $xy = c^2$
   (ii) $\dfrac{x^2}{a^2} - \dfrac{y^2}{b^2} = 1$
   (iii) $\dfrac{y^2}{a^2} - \dfrac{x^2}{b^2} = 1$

(iv) $x^2 + y^2 = r^2$

(v) $\dfrac{x^2}{a^2} + \dfrac{y^2}{b^2} = 1$

(vi) $xy + y^2 = 5$

(vii) $x^2 + y^2 - 3xy + 5y = 0$

(viii) $x^2 + y^2 + 2gx + 2fy + C = 0$

(ix) $y^2 = 4ax$

(x) $x^2 = -5y$.

    (a) with respect to x

    (b) with respect to y.

9. Determine the gradients at the points shown adjacent to each function.

    (i) $y = 3x^2 - 5x - 7$ $(x = -1)$

    (ii) $y = \dfrac{x}{x-1}$ $(x = 2)$

    (iii) $y = x(x^3 - 1)(x^2 + 1)$ $(x = 0)$

    (iv) $y = \dfrac{5}{x}$ $(x = 5)$

    (v) $y = -x^{-\frac{2}{3}}$ $(x = 1)$

    (vi) $y^2 = x$ $(x = 4)$

    (vii) $x^2 = -4y$ $(x = -2)$

    (viii) $y = \dfrac{1}{x} + \dfrac{2}{x^2} + \dfrac{3}{x^3}$ $(x = -1)$

    (ix) $x^2 + y^2 = 4$ $(x = 1)$

    (x) $xy - x^2 + y^2 - 1 = 0$ $(x = 0)$

10. Differentiate the following functions:-

    (i) $y = anx^{n-1}$

    (ii) $y = \dfrac{1}{\sqrt{x}} + \sqrt{x} + \sqrt[3]{x^2}$

    (iii) $y = \sqrt{x^2 - 1}\sqrt{x^2 + 1}$

    (iv) $y = \dfrac{x^2 - 1}{x + 1}$

(v) $y = \dfrac{x^3 - 1}{x^3 + 2}$

(vi) $y = (1 - 3x)^{\frac{1}{5}}$

(vii) $y = x^2\sqrt{x - 1}$

(viii) $xy = x^2 + y^2$

(ix) $3x^2 + 3y^2 = x - y - 5$

(x) $3x^2 + 5y^2 = 25$.

11. Determine the derivatives of the following functions with respect to x:-

    (i) $(2x + 5)^4$

    (ii) $(x - 1)^{-4}$

    (iii) $(1 + 3x)^{\frac{1}{2}}$

    (iv) $(1 + 2x + 3x^2)^5$

    (v) $(5x - 7)^{\frac{1}{2}}$

    (vi) $(3x - 2)^{-\frac{1}{2}}$

    (vii) $(1 + 7x)^{\frac{1}{2}}$

    (viii) $\dfrac{1}{\sqrt{1 + 4x^2}}$

    (ix) $(5x^2 + 7x - 3)^2$

    (x) $\dfrac{x^3 - 2x^2 - 7x + 1}{x^2 - 1}$.

12. Determine $\dfrac{dy}{dx}$ from the following equations:-

    (i) $\sqrt{x} + \sqrt{y} = \sqrt{3}$

    (ii) $2x^2 + 3xy = 5$

    (iii) $x^2 + y^2 = 3^2$

    (iv) $5y^2 = 4x$

    (v) $2x + 3y = \sqrt{7}$

    (vi) $2x^2 + 3y^2 = 25$.

13. For the functions in 12, find the numerical values of $\dfrac{dy}{dx}$ at $x = 1$.

14. Differentiate with respect to x:-

    (i) $\dfrac{x-7}{x^2-x+5}$

    (ii) $\dfrac{1}{1-x}$

    (iii) $\dfrac{\sqrt{x}+1}{\sqrt{x}}$

    (iv) $\dfrac{1}{2x^2+3x+4}$

    (v) $\dfrac{x-1}{\sqrt{x}}$.

15. Find the following:

    (i) $\dfrac{d}{dx}(5x^4+5x-1)$

    (ii) $\dfrac{d}{dy}\left(\dfrac{y-1}{y+1}\right)$

    (iii) $\dfrac{d}{dZ}\left(\dfrac{1}{Z}+\dfrac{1}{Z^2}+\dfrac{1}{Z^3}\right)$

    (iv) $\dfrac{d}{dt}\left(\dfrac{t^4+1}{2t}\right)$

    (v) $\dfrac{d}{du}\{(u^2+1)(u^3+2)\}$

16. Find the derived functions of the following:-

    (i) $y=\dfrac{1}{\sqrt{x}}-\dfrac{1}{x}+\dfrac{1}{x^2}$

    (ii) $y=(x+1)(x+2)(x+3)$

    (iii) $y=\dfrac{\sqrt{x}-1}{\sqrt{x}+1}$

    (iv) $y=\left(1-\dfrac{1}{x^2}\right)^5$

    (v) $xy+x^2y^2+x^3y^3=0$.

17. If $pv=100$,

    (i) determine $\dfrac{dp}{dv}$ when $v=25$

    (ii) determine $\dfrac{dv}{dp}$ when $v=2$.

2

Trigonometric or Circular Functions

Determine the Derivative of Sin x from First Principles

$y = \sin x$, $y + \delta y = \sin(x + \delta x)$. If x is increased to $x + \delta x$ then y is increased to $y + \delta y$, subtracting the two equations

$$\delta y = \sin(x + \delta x) - \sin x$$

$\delta y = \sin x \cos \delta x + \sin \delta x \cos x - \sin x$. As $\delta x \to 0$, $\cos \delta x \to 1$ and $\sin \delta x \to \delta x$ where x is expressed in radians

$$\delta y = \sin x + \delta x \cos x - \sin x \quad \delta y = \delta x \cos x$$

dividing each side by δx,

$$\frac{\delta y}{\delta x} = \cos x; \text{ as } \delta x \to 0, \quad \frac{\delta y}{\delta x} \to \frac{dy}{dx}$$

$$\boxed{\frac{dy}{dx} = \cos x}$$

Determine the Derivative of Cos x from First Principles

$y = \cos x$, $y + \delta y = \cos(x + \delta x)$, subtracting the two equations $\delta y = \cos(x + \delta x) - \cos x = \cos x \cos \delta x - \sin x \sin \delta x - \cos x$. As $\delta x \to 0$, $\cos \delta x \to 1$, $\sin \delta x \to \delta x$

$$\delta y = \cos x - \sin x \, \delta x - \cos x$$

$$\delta y = -\sin x \, \delta x$$

dividing each side by δx

$$\frac{\delta y}{\delta x} = -\sin x; \quad \text{as } \delta x \to 0, \quad \frac{\delta y}{\delta x} \to \frac{dy}{dx}$$

$$\boxed{\frac{dy}{dx} = -\sin x}$$

Therefore, the derivatives of $\sin x$ and $\cos x$ are respectively $\cos x$ and $-\sin x$.

$$y = \sin x \qquad \frac{dy}{dx} = \cos x$$

$$y = \cos x \qquad \frac{dy}{dx} = -\sin x.$$

The Derivative of Tan x

$$y = \tan x = \frac{\sin x}{\cos x}$$

$$\frac{dy}{dx} = \frac{\cos x \cos x - \sin x(-\sin x)}{\cos^2 x}$$

$$= \frac{\cos^2 x + \sin^2 x}{\cos^2 x} = \sec^2 x$$

$$y = \tan x \qquad \boxed{\frac{dy}{dx} = \sec^2 x}$$

The Derivative of Cot x

$$y = \cot x = \frac{\cos x}{\sin x}$$

$$\frac{dy}{dx} = \frac{-\sin x \sin x - \cos x \cos x}{\sin^2 x}$$

$$\frac{dy}{dx} = -\frac{\sin^2 x + \cos^2 x}{\sin^2 x}$$

$$\frac{dy}{dx} = -\csc^2 x.$$

$$y = \cot x \qquad \boxed{\frac{dy}{dx} = -\csc^2 x}$$

The Derivative of Cosec x

$$y = \operatorname{cosec} x = \frac{1}{\sin x}$$

$$\frac{dy}{dx} = -\frac{\cos x}{\sin^2 x} = -\frac{\cos x}{\sin x} \cdot \frac{1}{\sin x}$$

$$\frac{dy}{dx} = -\cot x \operatorname{cosec} x.$$

$$y = \operatorname{cosec} x \qquad \boxed{\frac{dy}{dx} = -\cot x \operatorname{cosec} x}$$

The Derivative of Sec x

$$y = \sec x = \frac{1}{\cos x}$$

$$\frac{dy}{dx} = \frac{\sin x}{\cos^2 x} = \frac{\sin x}{\cos x} \cdot \frac{1}{\cos x} = \tan x \sec x$$

$$y = \sec x \qquad \boxed{\frac{dy}{dx} = \tan x \sec x}$$

Function	Derivative
y	$\dfrac{dy}{dx}$
$\sin x$	$\cos x$
$\cos x$	$-\sin x$
$\tan x$	$\sec^2 x$
$\cot x$	$-\operatorname{cosec}^2 x$
$\operatorname{cosec} x$	$-\cot x \operatorname{cosec} x$
$\sec x$	$\tan x \sec x$

The table above shows the derivatives of the six basic trigonometric functions which must be learnt and written down.

The Derivatives of $\sin kx$, $\cos kx$, $\tan kx$, $\cot kx$, $\operatorname{cosec} kx$, $\sec kx$ Using Function of a Function.

Let $u = kx$, $\dfrac{du}{dx} = k$

$y = \sin kx = \sin u,$ $\dfrac{dy}{du} = \cos u$

$y = \cos kx = \cos u,$ $\dfrac{dy}{du} = -\sin u$

$y = \tan kx = \tan u,$ $\dfrac{dy}{du} = \sec^2 u$

$y = \cot kx = \cot u,$ $\dfrac{dy}{du} = -\operatorname{cosec}^2 u$

$y = \operatorname{cosec} kx = \operatorname{cosec} u,$ $\dfrac{dy}{du} = -\cot u \operatorname{cosec} u$

$y = \sec kx = \sec u,$ $\dfrac{dy}{du} = \tan u \sec u$

$y = \sin kx$ $\dfrac{dy}{dx} = k \cos kx$

$y = \cos kx$ $\dfrac{dy}{dx} = -k \sin kx$

$y = \tan kx$ $\dfrac{dy}{dx} = k \sec^2 kx$

$y = \cot kx$ $\dfrac{dy}{dx} = -k \operatorname{cosec}^2 kx$

$y = \operatorname{cosec} kx$ $\dfrac{dy}{dx} = -k \cot kx \operatorname{cosec} kx$

$y = \sec kx$ $\dfrac{dy}{dx} = k \tan kx \sec kx$

WORKED EXAMPLE 14

Differentiate the following functions:-

(i) $y = 3 \sin x$
(ii) $y = -\sin 3x$
(iii) $y = 2 \cos x$
(iv) $y = -4 \cos \dfrac{x}{2}$
(v) $y = \tan x$
(vi) $y = 3 \tan 3x$
(vii) $y = -\cot x$
(viii) $y = \cot 5x$
(ix) $y = \operatorname{cosec} \dfrac{x}{2}$
(x) $y = 3 \operatorname{cosec} 3x$
(xi) $y = \sec \dfrac{x}{3}$
(xii) $y = 3 \sec 4x$.

Solution 14

(i) $y = 3 \sin x$ $\dfrac{dy}{dx} = 3 \cos x$

(ii) $y = -\sin 3x$ $\dfrac{dy}{dx} = -3 \cos 3x$

Trigonometric or Circular Functions

(iii) $y = 2\cos x \qquad \dfrac{dy}{dx} = -2\sin x$

(iv) $y = -4\cos \dfrac{x}{2} \qquad \dfrac{dy}{dx} = 2\sin \dfrac{x}{2}$

(v) $y = \tan x \qquad \dfrac{dy}{dx} = \sec^2 x$

(vi) $y = 3\tan 3x \qquad \dfrac{dy}{dx} = 9\sec^2 3x$

(vii) $y = -\cot x \qquad \dfrac{dy}{dx} = \operatorname{cosec}^2 x$

(viii) $y = \cot 5x \qquad \dfrac{dy}{dx} = -5\operatorname{cosec}^2 5x$

(ix) $y = \operatorname{cosec} \dfrac{x}{2} \qquad \dfrac{dy}{dx} = -\dfrac{1}{2}\operatorname{cosec}\dfrac{x}{2}\cot\dfrac{x}{2}$

(x) $y = 3\operatorname{cosec} 3x \qquad \dfrac{dy}{dx} = -9\operatorname{cosec} 3x \cot 3x$

(xi) $y = \sec \dfrac{x}{3} \qquad \dfrac{dy}{dx} = \dfrac{1}{3}\sec\dfrac{x}{3}\tan\dfrac{x}{3}$

(xii) $y = 3\sec 4x \qquad \dfrac{dy}{dx} = 12\sec 4x \tan 4x$

(ii) $y = \sin x \tan x \qquad \dfrac{dy}{dx} = \cos x \tan x + \sin x \sec^2 x$

$\qquad\qquad = \sin x \left(1 + \sec^2 x\right)$

(iii) $y = \cos x \tan x = \cos x \dfrac{\sin x}{\cos x} = \sin x$

$\dfrac{dy}{dx} = \cos x$

(iv) $y = \cos x \cot x$

$\dfrac{dy}{dx} = -\sin x \cot x + \cos x \left(-\operatorname{cosec}^2 x\right)$

$\qquad = -\cos x \left(1 + \operatorname{cosec}^2 x\right)$

(v) $y = \operatorname{cosec} x \sec x$

$\dfrac{dy}{dx} = -\operatorname{cosec} x \cot x \sec x + \operatorname{cosec} x \tan x \sec x$

$\dfrac{dy}{dx} = -\dfrac{1}{\sin x} \cdot \dfrac{\cos x}{\sin x} \cdot \dfrac{1}{\cos x} + \dfrac{1}{\sin x} \cdot \dfrac{\sin x}{\cos x} \cdot \dfrac{1}{\cos x}$

$\qquad = -\operatorname{cosec}^2 x + \sec^2 x.$

Worked Example 15

Determine the derivatives of the following circular functions:

(i) $y = \sin x \cos x$

(ii) $y = \sin x \tan x$

(iii) $y = \cos x \tan x$

(iv) $y = \cos x \cot x$

(v) $y = \operatorname{cosec} x \sec x.$

Solution 15

(i) $y = \sin x \cos x,$ using the product rule

$\dfrac{dy}{dx} = \cos x \cos x + \sin x (-\sin x)$

$\qquad = \cos^2 x - \sin^2 x = \cos 2x,$ alternatively

$y = \sin x \cos x = \dfrac{\sin 2x}{2}$

$\dfrac{dy}{dx} = \dfrac{2}{2}\cos 2x = \cos 2x$

Worked Example 16

Determine the gradients of the following functions at the points indicated adjacent to each function:

(i) $y = 2\sin x \qquad \left(x = \dfrac{\pi}{2}\right)$

(ii) $y = 3\tan x \qquad \left(x = \dfrac{\pi}{4}\right)$

(iii) $y = 3\cos \dfrac{x}{2} \qquad (x = \pi)$

(iv) $y = -\cot x \qquad \left(x = \dfrac{\pi}{4}\right)$

(v) $y = \dfrac{\pi}{2}\sin 2x \qquad \left(x = \dfrac{\pi}{4}\right)$

(vi) $y = \operatorname{cosec} \dfrac{x}{2} \qquad (x = \pi)$

(vii) $y = 3\sec 3x \qquad \left(x = \dfrac{\pi}{6}\right)$

(viii) $y = 5\cos 5x \qquad \left(x = \dfrac{\pi}{2}\right)$

(ix) $y = 2\sin x \tan x \qquad \left(x = \dfrac{\pi}{4}\right)$

(x) $y = 5\cos x \cot x \qquad \left(x = \dfrac{3\pi}{2}\right).$

Solution 16

(i) $y = 2\sin x \quad \dfrac{dy}{dx} = 2\cos x = 2\cos\dfrac{\pi}{2} = 0$

(ii) $y = 3\tan x \quad \dfrac{dy}{dx} = 3\sec^2 x = 3\sec^2\dfrac{\pi}{4} = 6$

(iii) $y = 3\cos\dfrac{x}{2}$

$\dfrac{dy}{dx} = -\dfrac{3}{2}\sin\dfrac{x}{2} = -\dfrac{3}{2}\sin\dfrac{\pi}{2} = -\dfrac{3}{2}$

(iv) $y = -\cot x \quad \dfrac{dy}{dx} = \text{cosec}^2 x = \text{cosec}^2\dfrac{\pi}{4} = 2$

(v) $y = \dfrac{\pi}{2}\sin 2x$

$\dfrac{dy}{dx} = \dfrac{\pi}{2} \cdot 2\cos 2x = \pi\cos\dfrac{\pi}{2} = 0$

(vi) $y = \text{cosec}\dfrac{x}{2}$

$\dfrac{dy}{dx} = -\dfrac{1}{2}\text{cosec}\dfrac{x}{2}\cot\dfrac{x}{2} = -\dfrac{1}{2}\text{cosec}\dfrac{\pi}{2}\cot\dfrac{\pi}{2}$

$= 0$

(vii) $y = 3\sec 3x$

$\dfrac{dy}{dx} = 9\sec 3x \tan 3x = 9\sec 3\dfrac{\pi}{6}\tan 3\dfrac{\pi}{6} = \infty$

(viii) $y = 5\cos 5x$

$\dfrac{dy}{dx} = -25\sin 5x = -25\sin 5\dfrac{\pi}{2} = -25$

(ix) $y = 2\sin x \tan x$

$\dfrac{dy}{dx} = 2\cos x \tan x + 2\sin x \sec^2 x$

$\dfrac{dy}{dx} = 2\cos\dfrac{\pi}{4}\tan\dfrac{\pi}{4} + 2\sin\dfrac{\pi}{4}\sec^2\dfrac{\pi}{4}$

$= 2 \cdot \dfrac{1}{\sqrt{2}} \cdot 1 + 2 \cdot \dfrac{1}{\sqrt{2}} \cdot 2 = \dfrac{6}{\sqrt{2}}\dfrac{\sqrt{2}}{\sqrt{2}} = 3\sqrt{2}$

(x) $y = 5\cos x \cot x$

$\dfrac{dy}{dx} = -5\sin x \cot x - 5\cos x \, \text{cosec}^2 x$

$= -5\sin\dfrac{3\pi}{2}\cot\dfrac{3\pi}{2} - 5\cos\dfrac{3\pi}{2}\text{cosec}^2\dfrac{3\pi}{2}$

$= 0.$

Worked Example 17

Differentiate the following functions:-

(i) $y = 3\sin\left(x - \dfrac{\pi}{3}\right)$

(ii) $y = -5\cos(\alpha - x)$

(iii) $y = \cot\left(x + \dfrac{\pi}{2}\right)$

(iv) $y = 5\tan\left(3x - \dfrac{2\pi}{5}\right)$

(v) $y = \dfrac{5\sin x}{\cos\left(x - \dfrac{\pi}{2}\right)}$

(vi) $y = \tan 2x \sec \alpha$.

Solution 17

(i) $y = 3\sin\left(x - \dfrac{\pi}{3}\right)$

$\dfrac{dy}{dx} = 3\cos\left(x - \dfrac{\pi}{3}\right)$

(ii) $y = -5\cos(\alpha - x)$

$\dfrac{dy}{dx} = -\{-5\sin(\alpha - x)(-1)\}$

$= -5\sin(\alpha - x)$

(iii) $y = \cot\left(x + \dfrac{\pi}{2}\right)$

$\dfrac{dy}{dx} = -\text{cosec}^2\left(x + \dfrac{\pi}{2}\right)$

(iv) $y = 5\tan\left(3x - \dfrac{2\pi}{5}\right)$

$\dfrac{dy}{dx} = 15\sec^2\left(3x - \dfrac{2\pi}{5}\right)$

(v) $y = \dfrac{5\sin x}{\cos\left(x - \dfrac{\pi}{2}\right)}$

$\dfrac{dy}{dx} = 0$

(vi) $y = \tan 2x \sec \alpha$

$\dfrac{dy}{dx} = 2\sec \alpha \sec^2 2x$

Trigonometric or Circular Functions

Worked Example 18

Differentiate the following functions:-

(i) $y = x^2 \sin x$

(ii) $y = \sin^3 x$

(iii) $y = 5 \sin^2 x \cos^3 x$

Solution 18

(i) $y = x^2 \sin x \qquad \dfrac{dy}{dx} = 2x \sin x + x^2 \cos x$

(ii) $y = \sin^3 x \quad$ Let $u = \sin x \quad \dfrac{du}{dx} = \cos x$

$y = u^3 \qquad \dfrac{dy}{du} = 3u^2 \qquad \dfrac{dy}{dx} = 3 \cos x \sin^2 x$

(iii) $y = 5 \sin^2 x \cos^3 x$

$\dfrac{dy}{dx} = 10 \sin x \cos^4 x + 15 \sin^2 x (-\sin x) \cos^2 x$

$= 10 \sin x \cos^4 x - 15 \sin^3 x \cos^2 x.$

Implicit Functions

If $y \tan x = y^2 x^2 + 3$, determine the value of $\dfrac{dy}{dx}$.

Differentiating with respect to x

$\dfrac{dy}{dx} \tan x + y \sec^2 x = 2y \dfrac{dy}{dx} x^2 + y^2 2x$

$\dfrac{dy}{dx} \left(\tan x - 2yx^2 \right) = 2xy^2 - y \sec^2 x$

$\dfrac{dy}{dx} = \dfrac{y(2xy - \sec^2 x)}{\tan x - 2yx^2}.$

Function of a Function of a Function

$y = 5 \sin^3 (x^2 + 1)$, determine the value of $\dfrac{dy}{dx}$.

Let $u = x^2 + 1 \qquad \dfrac{du}{dx} = 2x$

$y = 5 \sin^3 u.$

Let $w = \sin u \qquad \dfrac{dw}{du} = \cos u$

$y = 5w^3 \qquad \dfrac{dy}{dw} = 15w^2$

$\dfrac{dy}{dx} = \dfrac{dy}{dw} \cdot \dfrac{dw}{du} \cdot \dfrac{du}{dx} = 15w^2 \cdot \cos u \cdot 2x$

$= 30x \sin^2 (x^2 + 1) \cos (x^2 + 1).$

Inverse Trigonometric Functions

$y = \sin^{-1} x \qquad\qquad y = \cos^{-1} x$

$x = \sin y \qquad\qquad x = \cos y$

$\dfrac{dx}{dy} = \cos y \qquad\qquad \dfrac{dx}{dy} = -\sin y$

$\dfrac{dy}{dx} = \dfrac{1}{\sqrt{1 - \sin^2 y}} \qquad \dfrac{dy}{dx} = -\dfrac{1}{\sqrt{1 - \cos^2 y}}$

$\boxed{\dfrac{dy}{dx} = \dfrac{1}{\sqrt{1 - x^2}}} \qquad \boxed{\dfrac{dy}{dx} = -\dfrac{1}{\sqrt{1 - x^2}}}$

$y = \tan^{-1} x \qquad\qquad y = \cot^{-1} x$

$x = \tan y \qquad\qquad x = \cot y$

$\dfrac{dx}{dy} = \sec^2 y \qquad\qquad \dfrac{dx}{dy} = -\csc^2 y$

$\dfrac{dy}{dx} = \dfrac{1}{1 + \tan^2 y} \qquad \dfrac{dy}{dx} = -\dfrac{1}{1 + \cot^2 y}$

$\boxed{\dfrac{dy}{dx} = \dfrac{1}{1 + x^2}} \qquad \boxed{\dfrac{dy}{dx} = -\dfrac{1}{1 + x^2}}$

$y = \sec^{-1} x \qquad\qquad y = \csc^{-1} x$

$x = \sec y \qquad\qquad x = \csc y$

$\dfrac{dx}{dy} = \sec y \tan y \qquad \dfrac{dx}{dy} = -\csc y \cot y$

$\dfrac{dy}{dx} = \dfrac{1}{\sec y \tan y} \qquad \dfrac{dy}{dx} = -\dfrac{1}{\csc y \cot y}$

$\dfrac{dy}{dx} = \dfrac{1}{x(\sec^2 y - 1)^{\frac{1}{2}}} \qquad \dfrac{dy}{dx} = -\dfrac{1}{x(\csc^2 y - 1)^{\frac{1}{2}}}$

$\boxed{\dfrac{dy}{dx} = \dfrac{1}{x(x^2 - 1)^{\frac{1}{2}}}} \qquad \boxed{\dfrac{dy}{dx} = -\dfrac{1}{x(x^2 - 1)^{\frac{1}{2}}}}$

Identities

$$\sin^2 y + \cos^2 y = 1$$

$$1 + \tan^2 y = \sec^2 y$$

$$1 + \cot^2 y = \csc^2 y$$

Function	Derivative
y	$\dfrac{dy}{dx}$
$\sin^{-1} x$	$\dfrac{1}{(1-x^2)^{\frac{1}{2}}}$
$\cos^{-1} x$	$-\dfrac{1}{(1-x^2)^{\frac{1}{2}}}$
$\tan^{-1} x$	$\dfrac{1}{1+x^2}$
$\cot^{-1} x$	$-\dfrac{1}{1+x^2}$
$\sec^{-1} x$	$\dfrac{1}{x(x^2-1)^{\frac{1}{2}}}$
$\csc^{-1} x$	$-\dfrac{1}{x(x^2-1)^{\frac{1}{2}}}$

Exercises 2

1. Differentiate from first principles the following circular functions:
   - (i) $y = 3 \sin x$
   - (ii) $y = -2 \cos x$
   - (iii) $y = \tan 2x$.

2. State the derivatives of the basic trigonometric functions:
   - (i) $y = \sin x$
   - (ii) $y = 2 \cos x$
   - (iii) $y = 3 \tan x$
   - (iv) $y = 4 \cot x$
   - (v) $y = 5 \csc x$
   - (vi) $y = 6 \sec x$.

3. Determine the gradients of the following trigonometric functions:-
   - (i) $y = \dfrac{1}{2} \sin 2x$
   - (ii) $y = 3 \cos \dfrac{x}{3}$
   - (iii) $y = 4 \tan \dfrac{x}{4}$
   - (iv) $y = \dfrac{1}{5} \csc 5x$
   - (v) $y = 7 \sec 7x$
   - (vi) $y = 2 \cot 3x$.

4. Evaluate the gradients of the functions of question 3 at the points:-
   - (a) $x = 0$
   - (b) $x = \dfrac{\pi}{4}$
   - (c) $x = \dfrac{3\pi}{4}$.

5. Differentiate with respect to x:
   - (i) $y = x \sin x$
   - (ii) $y = x^2 \sin^2 x$
   - (iii) $y = \tan^2 x$
   - (iv) $y = 3 \sec^2 x \tan x$
   - (v) $y = 5 \csc^3 x$
   - (vi) $y = \cot^4 x$.

6. Differentiate with respect to t:
   - (i) $y = 2 \sin^{\frac{3}{4}} t$
   - (ii) $y = -\tan^{\frac{5}{2}} 3t$
   - (iii) $y = \sqrt{\csc t}$.

7. Differentiate
   - (i) $x^2 \sqrt{\cos x}$
   - (ii) $x \sqrt{\sin x}$
   - (iii) $x \sqrt{\tan x}$, with respect to x.

8. If $xy = \tan y$, show that $\dfrac{dy}{dx} = \dfrac{y}{1 + x^2 y^2 - x}$.

9. If $y = \dfrac{\sqrt{1-x^2}}{\cos^{-1} x}$, find $\dfrac{dy}{dx}$.

10. If $y = \dfrac{\sin^{-1} x}{\sqrt{1-x^2}}$, find $\dfrac{dy}{dx}$ provided that $-1 < x < 1$.

11. Differentiate with respect to x $\quad y = \dfrac{\tan^{-1} x}{\cot^{-1} x}$

12. Find $\dfrac{dy}{dx}$ if $y = \dfrac{\csc^{-1} x}{\sec^{-1} x}$.

13. If $y = \dfrac{\cos^{-1} x}{\sin^{-1} x}$, find $\dfrac{dy}{dx}$.

14. Differentiate the following inverse trigonometric functions:-
    (i) $y = 3 \arcsin 3x$
    (ii) $y = -\arctan 2x$
    (iii) $y = 5 \arccos 4x$.

    NOTE: $\arcsin x = \sin^{-1} x$, $\arctan x = \tan^{-1} x$ and $\arccos x = \cos^{-1} x$.

15. Differentiate the following:-
    (i) $y = \sin x \cos^{-1} x$
    (ii) $y = 2 \cos x \sin^{-1} x$
    (iii) $y = 3 \tan x \cot^{-1} x$.

3

Exponential Functions

$$y = a^x$$

where the exponent x is the independent variable, a is a constant and y is the dependent variable.

If $a = 2$

$$y = 2^x$$

x	0	1	2	3	-1	-2	-3
y	2^0	2^1	2^2	2^3	2^{-1}	2^{-2}	2^{-3}
y	1	2	4	8	$\frac{1}{2}$	$\frac{1}{4}$	$\frac{1}{8}$

when x varies between -3 and $+3$ the curve looks like Fig. 4-I/5

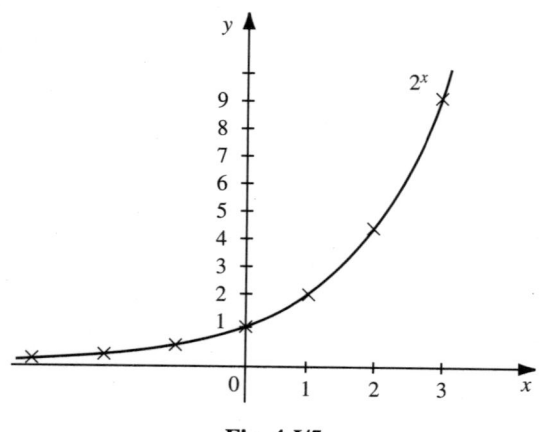

Fig. 4-I/5

y increases rapidly, abruptly or exponentially.

If $a = 3$

$$y = 3^x$$

x	-3	-2	-1	0	1	2	3
y	$\frac{1}{27}$	$\frac{1}{9}$	$\frac{1}{3}$	1	3	9	27

Plotting the values of y against x we have also an exponential graph as shown in Fig. 4-I/6

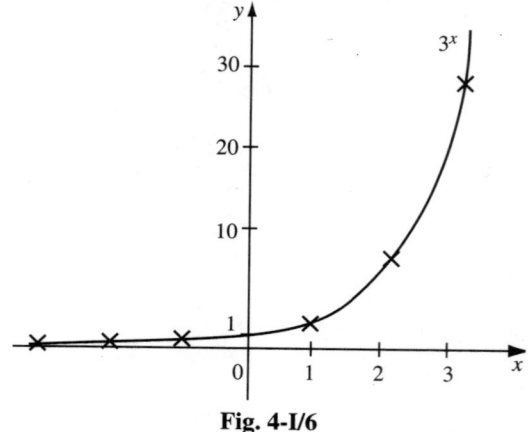

Fig. 4-I/6

y increases more abruptly than the previous expression $y = 2^x$.

There is a value between 2 and 3 such that the gradient of the function at a certain point is the same as the function; this is the only function in Mathematics whose gradient is the same as the function, $2 < e < 3$

$$y = e^x$$

$$e^x = 1 + \frac{x}{1!} + \frac{x^2}{2!} + \frac{x^3}{3!} + \frac{x^4}{4!} + \quad \ldots (1)$$

e^x is expressed as an infinite algebraic power series as shown above.

$$e^1 = 1 + \frac{1}{1!} + \frac{1^2}{2!} + \frac{1^3}{3!} + \ldots$$

$e = 2.718281828$ to ten significant figures.

Differentiating equation (1) with respect to x, we have

$$\frac{d}{dx}(e^x) = \frac{d}{dx}\left(1 + \frac{x}{1!} + \frac{x^2}{2!} + \frac{x^3}{3!} + \frac{x^4}{4!} + \ldots\right)$$

$$\frac{dy}{dx} = \frac{1}{1!} + \frac{2x}{2!} + \frac{3x^2}{3!} + \frac{4x^3}{4!} + \ldots$$

$$= 1 + \frac{x}{1!} + \frac{x^2}{2!} + \frac{x^3}{3!} + \ldots$$

$4! = 1 \times 2 \times 3 \times 4$ (factorial 4)

therefore $\boxed{\dfrac{dy}{dx} = e^x}$

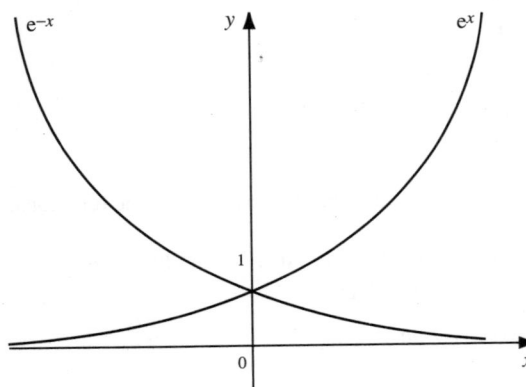

Fig. 4-I/7

Determine the Derivative of $y = e^x$ from First Principles

$y = e^x$, $y + \delta y = e^{x+\delta x}$

subtracting the equations

$$\delta y = e^{x+\delta x} - e^x = e^x \cdot e^{\delta x} - e^x = e^x\left(e^{\delta x} - 1\right)$$

$$= e^x\left(1 + \frac{\delta x}{1!} + \frac{\delta x^2}{2!} + \frac{\delta x^3}{3!} + \ldots - 1\right)$$

$$\delta y = e^x\left(\frac{\delta x}{1!} + \frac{\delta x^2}{2!} + \frac{\delta x^3}{3!} + \ldots\right)$$

dividing each side by δx

$$\frac{\delta y}{\delta x} = e^x\left(\frac{1}{1!} + \frac{\delta x}{2!} + \frac{\delta x^2}{3!} + \ldots\right)$$

as $\delta x \to 0$, $\dfrac{\delta y}{\delta x} \to \dfrac{dy}{dx}$

$$\boxed{\frac{dy}{dx} = e^x}$$

The Derivative of $y = e^{kx}$

Let $u = kx$, $\dfrac{du}{dx} = k$

$y = e^u$, $\dfrac{dy}{du} = e^u$

$$\frac{dy}{dx} = \frac{dy}{du}\frac{du}{dx} = e^u \cdot k = ke^{kx}$$

$$\boxed{\frac{dy}{dx} = ke^{kx}}$$

WORKED EXAMPLE 19

Differentiate the following exponential functions:-

(i) $y = 3e^x$

(ii) $y = 3e^{\frac{x}{3}}$

(iii) $y = e^{-3x}$

(iv) $y = 5e^{-5x}$

(v) $y = 2e^{x^2}$

(vi) $y = e^{-x^2}$

(vii) $y = e^{nx}$.

Solution 19

(i) $y = 3e^x$ $\dfrac{dy}{dx} = 3e^x$

(ii) $y = 3e^{\frac{x}{3}}$ $\dfrac{dy}{dx} = 3\left(\dfrac{1}{3}\right)e^{\frac{x}{3}} = e^{\frac{x}{3}}$

(iii) $y = e^{-3x}$ $\dfrac{dy}{dx} = -3e^{-3x}$

(iv) $y = 5e^{-5x}$ $\dfrac{dy}{dx} = -25e^{-5x}$

(v) $y = 2e^{x^2}$ let $u = x^2$, $\dfrac{du}{dx} = 2x$

$y = 2e^u$

$\dfrac{dy}{du} = 2e^u$

$\dfrac{dy}{dx} = 2x \cdot 2e^{x^2}$

$\dfrac{dy}{dx} = 4xe^{x^2}$

(vi) $y = e^{-x^2}$

$\dfrac{dy}{dx} = -2xe^{-x^2}$

(vii) $y = e^{nx}$

$\dfrac{dy}{dx} = ne^{nx}.$

(v) $y = e^{x^2} \cos x$

$\dfrac{dy}{dx} = 2xe^{x^2} \cdot \cos x + e^{x^2}(-\sin x)$

$\dfrac{dy}{dx} = 2xe^{x^2} \cos x - e^{x^2} \sin x$

(vi) $y = e^{-x} \tan x$

$\dfrac{dy}{dx} = -e^{-x} \tan x + e^{-x} \sec^2 x$

$\dfrac{dy}{dx} = e^{-x}\left(\sec^2 x - \tan x\right)$

WORKED EXAMPLE 20

Differentiate the following functions:-

(i) $y = xe^x$

(ii) $y = e^x \sin x$

(iii) $y = xe^{x^2}$

(iv) $y = x^2 e^{x^2}$

(v) $y = e^{x^2} \cos x$

(vi) $y = e^{-x} \tan x$.

Solution 20

(i) $y = xe^x$

$\dfrac{dy}{dx} = e^x + xe^x$

(ii) $y = e^x \sin x$

$\dfrac{dy}{dx} = e^x \sin x + e^x \cos x$

(iii) $y = xe^{x^2}$

$\dfrac{dy}{dx} = e^{x^2} + x \cdot 2xe^{x^2}$

$= e^{x^2} + 2x^2 e^{x^2}$

(iv) $y = x^2 e^{x^2}$

$\dfrac{dy}{dx} = 2x \cdot e^{x^2} + x^2 \cdot 2xe^{x^2}$

$= 2xe^{x^2} + 2x^3 e^{x^2}$

Exercises 3

1. Write down the power series of

    (i) e^x

    (ii) e^{-x}

    (iii) e^{2x}

    (iv) e^{-3x}

    and hence determine the derivative of the functions.

2. Write down the derivatives of the following exponential functions:-

    (i) $y = 3e^x$

    (ii) $y = e^{-3x}$

    (iii) $y = e^{x^2}$

    (iv) $y = e^{-3x^2}$

    (v) $y = ne^{ax}$

    (vi) $y = e^{\frac{x}{2}}$

    (vii) $y = \dfrac{1}{2}e^{-\frac{x}{2}}.$

3. Differentiate the following products:-

    (i) $y = 3x^2 e^{x^2}$

    (ii) $y = e^x \sin x$

    (iii) $y = e^{-3x} \cos 3x$

    (iv) $y = 3e^{-x} \sec x$

    (v) $y = e^{3x}\left(x^3 + 3\right).$

4. Determine the gradients of the quotients:-

   (i) $y = \dfrac{e^x}{\sin x}$

   (ii) $y = \dfrac{\tan x}{e^{2x}}$

   (iii) $y = \dfrac{e^{3x} \sin x}{\cot x}$.

5. Differentiate the following functions with respect to x:-

   (i) $y = x^3 e^{3x}$

   (ii) $y = e^{-2x}(x^2 - 1)$

   (iii) $y = 3e^{2x} \cos 2x$

   (iv) $y = \dfrac{e^x}{\tan x}$

   (v) $y = \sin(e^{3x})$

   (vi) $y = 3e^{x^2}$

   (vii) $y = e^{x^2} \sin x^2$

   (viii) $y = 3(e^{3x})^5$

   (ix) $y = e^x - e^{-x}$

   (x) $y = e^{2x} + e^{-2x}$

   (xi) $y = \dfrac{e^x + e^{-x}}{e^x - e^{-x}}$

   (xii) $y = e^{e^{-x}}$

   (xiii) $y = (e^{2x} + e^{-2x})^{-3}$

   (xiv) $y = 3a^x$

   (xv) $y = 2(3^x)$.

6. Differentiate the following functions:-

   (i) $y = e^{nx} \sin kx$

   (ii) $y = e^{-mx} \cos nx$

   (iii) $y = \dfrac{e^{ax}}{\cos bx}$

   (iv) $y = e^{2x} \cos 5x$

   (v) $y = e^x \sin 5x$.

7. Differentiate the following functions:-

   (i) $y = e^{-\frac{1}{t}}$

   (ii) $y = e^{\sqrt{u}}$

   (iii) $y = e^{-\sin x}$.

8. Differentiate from first principles the following:-

   (i) $y = e^{-x}$

   (ii) $y = e^{2x}$.

9. Sketch the graphs:-

   (i) $y = 2e^x$

   (ii) $y = 3e^{-x}$.

10. Differentiate from first principles $y = e^{\frac{1}{x}}$.

11. Sketch the curve $y = e^x \sin x$ for $0 \leq x \leq 4\pi$.

12. Sketch the curve $y = e^{-x} \sin x$ when $0 \leq x \leq 4\pi$.

13. Given that $y = e^{2x} \sin 3x$, determine $\dfrac{dy}{dx}$ and $\dfrac{d^2y}{dx^2}$ and hence express $\dfrac{d^2y}{dx^2}$ in the form $R e^{2x} \cos(3x+\alpha)$ giving values of R and $\tan \alpha$.

14. If $y = e^x \sin x$, show that $3\dfrac{d^2y}{dx^2} - 6\dfrac{dy}{dx} + 6y = 0$.

15. If $y = e^{-2x} \cos 2x$, show that $\dfrac{d^2y}{dx^2} + 4\dfrac{dy}{dx} + 8y = 0$.

16. Differentiate with respect to x:-

    (i) $\dfrac{e^{x^2}}{\cos x}$

    (ii) $e^{-x^2} \sin x$

    (iii) $(e^x \tan 2x)^3$.

17. Differentiate with respect to x:-

    (i) $e^{-\frac{1}{x}} \sin x$

    (ii) $e^{-\frac{1}{x^2}} \cos 2x$.

18. If $e^{2x} = \cos 3y$, find $\dfrac{dy}{dx}$

19. If $e^{\sqrt{y}} = \tan x$, find $\dfrac{dy}{dx}$.

20. If $\tan^3 y = e^{-2x}$, find $\dfrac{dy}{dx}$.

21. Differentiate implicitly

    $y \tan x = x^2$ and determine $\dfrac{d^2y}{dx^2}$.

22. Differentiate explicitly

    $y \tan x = x^2$ and determine $\dfrac{d^2y}{dx^2}$.

Logarithmic Functions

$y = \log_e x$ where $x > 0$,

by definition of the logarithm

$$e^y = x$$

$$x = e^y$$

$$\frac{dx}{dy} = e^y$$

$$\frac{dy}{dx} = \frac{1}{e^y} = \frac{1}{x}$$

$$\boxed{\frac{dy}{dx} = \frac{1}{x}}$$

The derivative of $\log_e x$ is $\frac{1}{x}$. $\log_e x$ can be written as $\ln x$

$$\log_e x = \ln x$$

If $y = \ln x$, $\frac{dy}{dx} = \frac{1}{x}$.

The graph of $y = \ln x$ is shown in Fig. 4-I/8

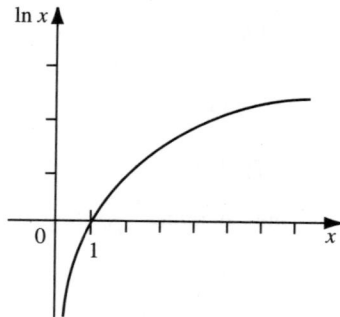

Fig. 4-I/8

To Determine the Derivative of $y = a^x$

$$y = a^x$$

$$\log_e y = x \log_e a$$

$$\frac{d}{dx}(\ln y) = \frac{d}{dx}(x \ln a)$$

$$\frac{1}{y}\frac{dy}{dx} = \ln a$$

$$\frac{dy}{dx} = y \ln a$$

$$\boxed{\frac{dy}{dx} = a^x \ln a}$$

Worked Example 21

Determine the derivatives of the functions:-

(i) $y = 2^x$

(ii) $y = 3^x$

(iii) $y = a^x$

(iv) $y = 5^x$

(v) $y = 10^x$

Solution 21

(i) $y = 2^x$, $\quad \frac{dy}{dx} = 2^x \ln 2$

(ii) $y = 3^x$, $\quad \frac{dy}{dx} = 3^x \ln 3$

(iii) $y = a^x$, $\quad \frac{dy}{dx} = a^x \ln a$

(iv) $y = 5^x$, $\quad \dfrac{dy}{dx} = 5^x \ln 5$

(v) $y = 10^x$, $\quad \dfrac{dy}{dx} = 10^x \ln 10$.

The Derivative of $y = \ln nx$ where n is a Positive Constant and $x > 0$

$y = \ln nx$

let $u = nx$, $\quad \dfrac{du}{dx} = n$

$y = \ln u$

$\dfrac{dy}{du} = \dfrac{1}{u}$

$\dfrac{dy}{dx} = \dfrac{dy}{du} \dfrac{du}{dx} = \dfrac{1}{u} \cdot n = \dfrac{1}{nx} \cdot n = \dfrac{1}{x}$

$$\boxed{\dfrac{d}{dx}(\ln nx) = \dfrac{1}{x}.}$$

Alternatively,

$y = \ln nx = \ln n + \ln x$

$\dfrac{dy}{dx} = 0 + \dfrac{1}{x} = \dfrac{1}{x}$

$\dfrac{d}{dx}(\ln nx) = \dfrac{1}{x}.$

Referring to Fig. 4-I/8, it is observed that the gradient is always positive since $x > 0$.

WORKED EXAMPLE 22

Determine the derivatives of the following logarithmic functions:-

(i) $y = 5\log_e x$,
(ii) $y = \log_{10} x$
(iii) $y = \log_e 3x$
(iv) $y = x \log_e x^2$
(v) $y = x^2 \ln x$
(vi) $y = x \ln x - x$.

Solution 22

(i) $y = 5\log_e x$,

$\dfrac{dy}{dx} = \dfrac{5}{x}$

(ii) $y = \log_{10} x = \dfrac{\log_e x}{\log_e 10}$

$\dfrac{dy}{dx} = \dfrac{1}{\ln 10} \cdot \dfrac{1}{x} = \dfrac{1}{x \ln 10}$

(iii) $y = \ln 3x$,

$\dfrac{dy}{dx} = 3 \cdot \dfrac{1}{3x} = \dfrac{1}{x}$

(iv) $y = x \log_e x^2$

$\dfrac{dy}{dx} = \ln x^2 + x \cdot \dfrac{2x}{x^2}$,

using the product rule. $\quad \dfrac{dy}{dx} = \ln x^2 + 2$

(v) $y = x^2 \ln x$

$\dfrac{dy}{dx} = 2x \cdot \ln x + x^2 \dfrac{1}{x} = 2x \ln x + x$

(vi) $y = x \ln x - x$

$\dfrac{dy}{dx} = 1 \cdot \ln x + x \cdot \dfrac{1}{x} - 1 = \ln x$

$\dfrac{dy}{dx} = \ln x$.

Exercises 4

1. Differentiate with respect to x the following:-

    (i) $y = \log|x|$
    (ii) $y = 3\ln|x|$
    (iii) $y = (3x)^x$
    (iv) $y = 7^x$
    (v) $y = \ln|kx|$.

2. Differentiate the following functions with respect to x:

    (i) $y = (\cos x)^x$
    (ii) $y = (\cot x)^x$

(iii) $y = (x+1)^x$

(iv) $y = \sqrt{\dfrac{(x^2+1)}{(x^3-1)(x^4+1)}}$.

(v) $y = \sqrt[3]{\dfrac{(x-1)}{(x+1)(x+2)}}$.

3. Determine the gradients of the functions:-

   (i) $y = \ln 5x^{\frac{1}{3}}$

   (ii) $y = \ln \left|\dfrac{1-x}{x}\right|$

   (iii) $y = x^2 \ln x$.

4. Find $\dfrac{dy}{dx}$ for the following:-

   (i) $y = x^{-3} \ln 3x$

   (ii) $y = x - \ln x$

   (iii) $y = \dfrac{x}{\ln x}$

(iv) $y = \dfrac{\ln 2x}{\sin 2x}$

(v) $y = \dfrac{\tan x}{\ln \left|\dfrac{1}{x}\right|}$.

5. If $y = e^x \ln x$, determine $\dfrac{dy}{dx}$.

6. If $y = e^{\sin x} \cos(\ln x)$, determine $\dfrac{dy}{dx}$.

7. If $y = e^{\cos x} \ln(\sin x)$, determine $\dfrac{dy}{dx}$.

8. Differentiate $y = \log_e \sec^2(5x-1)$.

9. Determine the first derivatives of the following:-

   (i) $y = \sqrt{x(x-1)(x+2)}$

   (ii) $y = \sqrt{(x+1)(x+3)(x+4)}$.

10. If $y = \sqrt{\dfrac{3x^3-4}{5x^3+7}}$, find $\dfrac{dy}{dx}$

5

Hyperbolic Functions

$y = \sinh x = \dfrac{e^x - e^{-x}}{2}$

$\dfrac{dy}{dx} = \dfrac{e^x + e^{-x}}{2} = \cosh x$

$y = \sinh x \qquad \boxed{\dfrac{dy}{dx} = \cosh x}$

$y = \cosh x = \dfrac{e^x + e^{-x}}{2}$

$\dfrac{dy}{dx} = \dfrac{e^x - e^{-x}}{2} = \sinh x$

$y = \cosh x \qquad \boxed{\dfrac{dy}{dx} = \sinh x}$

$y = \tanh x = \dfrac{\sinh x}{\cosh x}$

$\dfrac{dy}{dx} = \dfrac{\cosh^2 x - \sinh x \sinh x}{\cosh^2 x}$

$= \dfrac{\cosh^2 x - \sinh^2 x}{\cosh^2 x} = \dfrac{1}{\cosh^2 x}$

$= \operatorname{sech}^2 x$, where $\cosh^2 x - \sinh^2 x = 1$

$y = \tanh x \qquad \boxed{\dfrac{dy}{dx} = \operatorname{sech}^2 x}$

$y = \operatorname{sech} x = \dfrac{1}{\cosh x}$

$\dfrac{dy}{dx} = \dfrac{0 \cdot \cosh x - 1 \cdot \sinh x}{\cosh^2 x} = -\dfrac{\sinh x}{\cosh^2 x}$

$= -\dfrac{\sinh x}{\cosh x} \dfrac{1}{\cosh x} = -\tanh x \operatorname{sech} x$

$y = \operatorname{sech} x \qquad \boxed{\dfrac{dy}{dx} = -\tanh x \operatorname{sech} x}$

$y = \coth x = \dfrac{\cosh x}{\sinh x}$

$\dfrac{dy}{dx} = \dfrac{\sinh x \sinh x - \cosh x \cosh x}{\sin^2 x}$

$= \dfrac{\sinh^2 x - \cosh^2 x}{\sinh^2 x} = -\dfrac{1}{\sinh^2 x}$

$= -\operatorname{cosech}^2 x$, where $\cosh^2 x - \sinh^2 x = 1$

$y = \coth x \qquad \boxed{\dfrac{dy}{dx} = -\operatorname{cosech}^2 x}$

$y = \operatorname{cosech} x = \dfrac{1}{\sinh x}$

$\dfrac{dy}{dx} = \dfrac{0 \sinh x - 1 \cosh x}{\sinh^2 x} = -\dfrac{\cosh x}{\sinh^2 x}$

$= -\dfrac{\cosh x}{\sinh x} \cdot \dfrac{1}{\sinh x} = -\coth x \operatorname{cosech} x$

$y = \operatorname{cosech} x \qquad \boxed{\dfrac{dy}{dx} = -\coth x \operatorname{cosech} x}$

Function	Derivative
y	$\dfrac{dy}{dx}$
$\sinh x$	$\cosh x$
$\cosh x$	$\sinh x$
$\tanh x$	$\operatorname{sech}^2 x$
$\coth x$	$-\operatorname{cosech}^2 x$
$\operatorname{cosech} x$	$-\coth x \operatorname{cosech} x$
$\operatorname{sech} x$	$-\tanh x \operatorname{sech} x$

Worked Example 23

Write down the derivatives of $\sinh x$, $\cosh x$, $\tanh x$, $\coth x$, $\operatorname{cosech} x$, and $\operatorname{sech} x$.

Solution 23

$\dfrac{d}{dx}(\sinh x) = \cosh x$,

$\dfrac{d}{dx}(\cosh x) = \sinh x$,

$\dfrac{d}{dx}(\tanh x) = \operatorname{sech}^2 x$,

$\dfrac{d}{dx}(\coth x) = -\operatorname{cosech}^2 x$,

$\dfrac{d}{dx}(\operatorname{cosech} x) = -\coth x \operatorname{cosech} x$,

$\dfrac{d}{dx}(\operatorname{sech} x) = -\tanh x \operatorname{sech} x$.

Function of a Function

$\dfrac{d}{dx}(\sinh kx) = k \cosh kx$

$\dfrac{d}{dx}(\cosh kx) = k \sinh kx$

$\dfrac{d}{dx}(\tanh kx) = k \operatorname{sech}^2 kx$

$\dfrac{d}{dx}(\coth kx) = -k \operatorname{cosech}^2 kx$

$\dfrac{d}{dx}(\operatorname{cosech} kx) = -k \coth kx \operatorname{cosech} kx$

$\dfrac{d}{dx}(\operatorname{sech} kx) = -k \tanh kx \operatorname{sech} kx$.

Worked Example 24

Determine the derivatives of the following hyperbolic functions:-

(i) $y = 3 \sinh 2x$

(ii) $y = -5 \cosh 3x$

(iii) $y = \tanh \dfrac{3x}{2}$

(iv) $y = 4 \coth 4x$

(v) $y = 2 \operatorname{cosech} 2x$

(vi) $y = 5 \operatorname{sech} 5x$.

Solution 24

(i) $y = 3 \sinh 2x$ $\qquad \dfrac{dy}{dx} = 6 \cosh 2x$

(ii) $y = -5 \cosh 3x$ $\qquad \dfrac{dy}{dx} = -15 \sinh 3x$

(iii) $y = \tanh \dfrac{3x}{2}$ $\qquad \dfrac{dy}{dx} = \dfrac{3}{2} \operatorname{sech}^2 \dfrac{3x}{2}$

(iv) $y = 4 \coth 4x$ $\qquad \dfrac{dy}{dx} = -16 \operatorname{cosech}^2 4x$

(v) $y = 2 \operatorname{cosech} 2x$ $\qquad \dfrac{dy}{dx} = -4 \operatorname{cosech} 2x \coth 2x$

(vi) $y = 5 \operatorname{sech} 5x$ $\qquad \dfrac{dy}{dx} = -25 \operatorname{sech} 5x \tanh 5x$.

Worked Example 25

Determine the derivatives of the following functions:-

(i) $y = \operatorname{sech}^2 x$

(ii) $y = \tanh^3 x$

(iii) $y = \cosh^5 x$

(iv) $y = \sinh^4 x$

(v) $y = \sqrt{2} \operatorname{cosech}^{\frac{1}{2}} x$

(vi) $y = \sqrt{3} \coth^{\frac{3}{4}} x$.

Solution 25

(i) $y = \operatorname{sech}^2 x$

let $u = \operatorname{sech} x \qquad y = u^2$

$\dfrac{du}{dx} = -\operatorname{sech} x \tanh x$

$\dfrac{dy}{du} = 2u$

$\dfrac{dy}{dx} = \dfrac{dy}{du} \cdot \dfrac{du}{dx} = -2 \operatorname{sech}^2 x \tanh x$

(ii) $y = \tanh^3 x$

let $u = \tanh x \qquad y = u^3$

$\dfrac{du}{dx} = \operatorname{sech}^2 x$

$\dfrac{dy}{du} = 3u^2$

$\dfrac{dy}{dx} = \dfrac{dy}{du} \cdot \dfrac{du}{dx} = 3 \tanh^2 x \operatorname{sech}^2 x$

(iii) $y = \cosh^5 x$

let $u = \cosh x$

$y = u^5$

$\dfrac{du}{dx} = \sinh x$

$\dfrac{dy}{du} = 5u^4$

$\dfrac{dy}{dx} = 5\cosh^4 x \sinh x$

(iv) $y = \sinh^4 x$

let $u = \sinh x$

$\dfrac{du}{dx} = \cosh x$

$y = u^4$

$\dfrac{dy}{du} = 4u^3$

$\dfrac{dy}{dx} = 4\sinh^3 x \cosh x$

(v) $y = \sqrt{2}\,\operatorname{cosech}^{\frac{1}{2}} x$,

let $u = \operatorname{cosech} x$

$y = \sqrt{2}\,u^{\frac{1}{2}}$

$\dfrac{du}{dx} = -\operatorname{cosech} x \coth x$

$\dfrac{dy}{du} = \dfrac{1}{2}\sqrt{2}\,u^{-\frac{1}{2}}$

$\dfrac{dy}{dx} = -\dfrac{1}{2}\sqrt{2}\,\dfrac{\operatorname{cosech} x \coth x}{\operatorname{cosech}^{\frac{1}{2}} x}$

$\qquad = -\dfrac{1}{2}\sqrt{2}\,\operatorname{cosech}^{\frac{1}{2}} x \coth x.$

(vi) $y = \sqrt{3}\,\coth^{\frac{3}{4}} x$,

let $u = \coth x$

$y = \sqrt{3}\,u^{\frac{3}{4}}$

$\dfrac{du}{dx} = -\operatorname{cosech}^2 x$

$\dfrac{dy}{du} = \dfrac{3\sqrt{3}}{4} u^{-\frac{1}{4}}$,

$\dfrac{dy}{dx} = -\dfrac{3\sqrt{3}}{4}\dfrac{\operatorname{cosech}^2 x}{\coth^{\frac{1}{4}} x}.$

Inverse Hyperbolic Functions

$y = \sinh^{-1} x$

$y = \cosh^{-1} x$

$x = \sinh y$

$x = \cosh y$

$\dfrac{dx}{dy} = \cosh y$

$\dfrac{dx}{dy} = \sinh y$

$\dfrac{dy}{dx} = \dfrac{1}{\cosh y} = \dfrac{1}{\sqrt{1+\sinh^2 y}}$

$\boxed{\quad = \dfrac{1}{\sqrt{1+x^2}} \quad}$

$\dfrac{dy}{dx} = \dfrac{1}{\sinh y} = \dfrac{1}{\sqrt{\cosh^2 y - 1}}$

$\boxed{\quad = \dfrac{1}{\sqrt{x^2-1}} \quad}$

$y = \tanh^{-1} x$

$y = \coth^{-1} x$

$x = \tanh y$

$x = \coth y$

$\dfrac{dx}{dy} = \operatorname{sech}^2 y$

$\dfrac{dx}{dy} = -\operatorname{cosech}^2 y$

$\dfrac{dy}{dx} = \dfrac{1}{\operatorname{sech}^2 y} = \dfrac{1}{1-\tanh^2 y}$

$\boxed{\quad = \dfrac{1}{1-x^2} \quad}$

$\dfrac{dy}{dx} = \dfrac{1}{-\operatorname{cosech}^2 y} = \dfrac{1}{1-\coth^2 y}$

$\boxed{\quad = \dfrac{1}{1-x^2} \quad}$

$y = \operatorname{sech}^{-1} x$

$y = \operatorname{cosech}^{-1} x$

$x = \operatorname{sech} y$

$x = \operatorname{cosech} y$

$\dfrac{\mathrm{d}x}{\mathrm{d}y} = -\operatorname{sech} y \tanh y$

$\dfrac{\mathrm{d}x}{\mathrm{d}y} = -\coth y \operatorname{cosech} y$

$\dfrac{\mathrm{d}y}{\mathrm{d}x} = -\dfrac{1}{\operatorname{sech} y \tanh y} = -\dfrac{1}{x(1 - \operatorname{sech}^2 y)^{\frac{1}{2}}}$

$$\boxed{= -\dfrac{1}{x\left(1 - x^2\right)^{\frac{1}{2}}}}$$

$\dfrac{\mathrm{d}y}{\mathrm{d}x} = \dfrac{1}{-\coth y \operatorname{cosech} y} = \dfrac{1}{-(1 + \operatorname{cosech}^2 y)^{\frac{1}{2}} x}$

$$\boxed{= -\dfrac{1}{\left(1 + x^2\right)^{\frac{1}{2}} x}}$$

Hyperbolic identities

$\cosh^2 x - \sinh^2 x = 1$

$1 - \tanh^2 x = \operatorname{sech}^2 x$

$1 - \coth^2 x = -\operatorname{cosech}^2 x.$

Function	Derivative
y	$\dfrac{\mathrm{d}y}{\mathrm{d}x}$
$\sinh^{-1} x$	$\dfrac{1}{\left(1 + x^2\right)^{\frac{1}{2}}}$
$\cosh^{-1} x$	$\dfrac{1}{\left(x^2 - 1\right)^{\frac{1}{2}}}$
$\tanh^{-1} x$	$\dfrac{1}{1 - x^2}$
$\coth^{-1} x$	$\dfrac{1}{1 - x^2}$
$\operatorname{sech}^{-1} x$	$-\dfrac{1}{x\left(1 - x^2\right)^{\frac{1}{2}}}$
$\operatorname{cosech}^{-1} x$	$-\dfrac{1}{x\left(1 + x^2\right)^{\frac{1}{2}}}$

The graphs of hyperbolic functions.

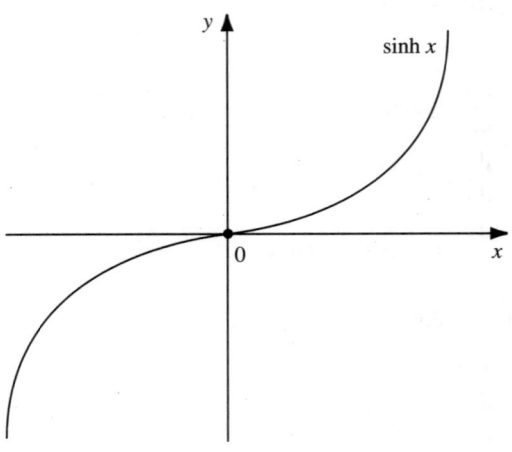

Fig. 4-I/9

The graphs of inverse hyperbolic functions

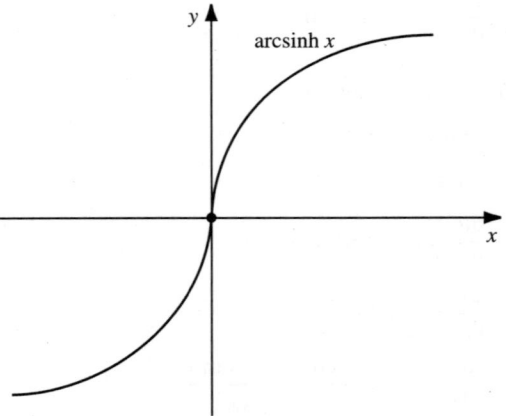

Fig. 4-I/10

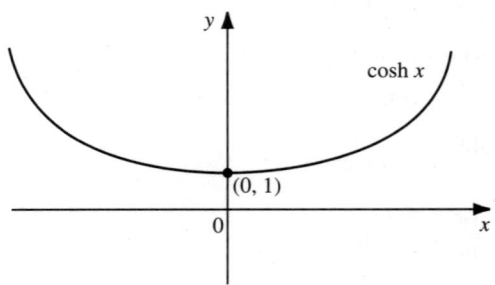

Fig. 4-I/11

Exercises 5

1. Derive the derivatives from first principles for the following hyperbolic functions:-
   (i) $y = \sinh x$
   (ii) $y = \cosh x$
   (iii) $y = \sinh \dfrac{1}{2}x$.

2. If $y = (\sinh^{-1} 3x)^2$, show that
$$(1 + 9x^2)\left(\dfrac{dy}{dx}\right)^2 = 36y.$$

3. Differentiate the following:-
   (i) $y = \tan 2x \coth 3x$
   (ii) $y = \sinh 3x \cot 2x$
   (iii) $y = \operatorname{cosech} \dfrac{1}{x}$
   (iv) $y = \operatorname{sech} x^2$
   (v) $y = 3\sinh^5 \dfrac{x}{2}$
   (vi) $y = \coth^{\frac{1}{2}} x \, \sinh^{\frac{3}{2}} x$.

4. Differentiate the following inverse hyperbolic functions:-
   (i) $y = 3\sinh^{-1}\dfrac{1}{x}$
   (ii) $y = \cosh^{-1} x^2$
   (iii) $y = 5\cosh^{-1}(x^2 - 3x + 2)$
   (iv) $y = \operatorname{cosech}^2 \dfrac{x}{2} \operatorname{sech}^2 \dfrac{x}{3}$.

5. If $y = \operatorname{sech}^{-1} 2x$, find $\dfrac{dy}{dx}$ and $\dfrac{d^2 y}{dx^2}$.

6. If $y = \sinh^{-1}\dfrac{1}{2}x$, find $\dfrac{d^2 y}{dx^2}$.

7. Show that $\dfrac{d}{dx}\left(\operatorname{cosech}^{-1} 2x\right)$
$$= -\dfrac{1}{x(1 + 4x^2)^{\frac{1}{2}}}.$$

8. Show that $\dfrac{d}{dx}\left(\tanh^{-1} 3x\right) = \dfrac{3}{(1 - 9x^2)}$.

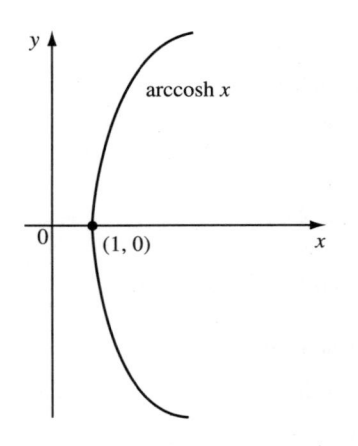

Fig. 4-I/12

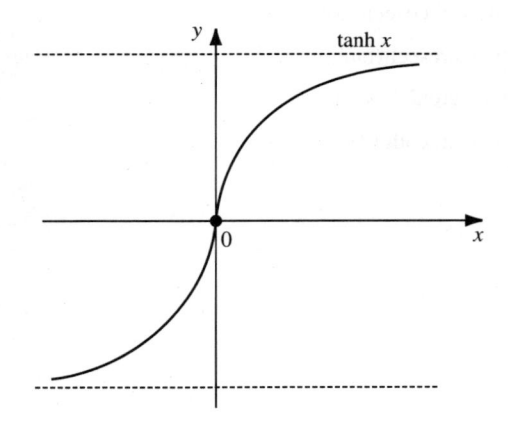

Fig. 4-I/13

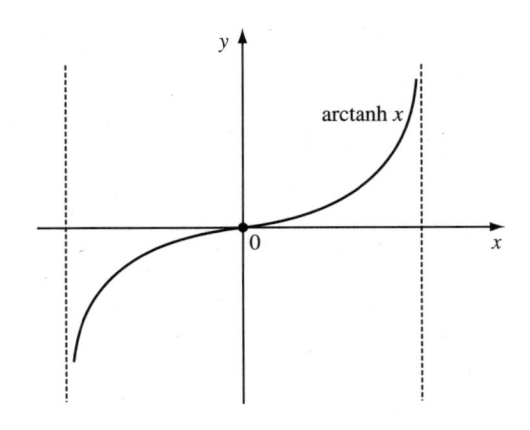

Fig. 4-I/14

9. Find the gradients of the functions
    (i) $y = \sinh x$
    (ii) $y = \cosh x$
    (iii) $y = \tanh x$
    at (a) $x = -5$ (b) $x = 0$ and (c) $x = 5$. Sketch the graphs and indicate these gradients.

10. Repeat (9) for
    (i) $y = \operatorname{arcsinh} x$
    (ii) $y = \operatorname{arccosh} x$
    (iii) $y = \operatorname{arctanh} x$ at
        (a) $x = 2$
        (b) $x = 1$.

11. Differentiate the following functions with respect to x:
    (i) $\sinh 2x \operatorname{cosech} 3x$
    (ii) $\sinh 3x$
    (iii) $e^x \cosh 2x$
    (iv) $\ln \sinh 5x$
    (v) $e^{\coth^2 x}$
    (vi) $x^3 \coth^3 5x$.
    (vii) $\sqrt{\coth 3x}$
    (viii) $\dfrac{1}{3} \cosh^3 x - \cosh x$
    (ix) $2 \tanh x \operatorname{sech}^2 x$
    (x) $\sqrt{\dfrac{\cosh 2x + 1}{\cosh 2x - 1}}$.

12. Differentiate:-
    (i) $\operatorname{arctanh}(\cosh 3x)$
    (ii) $\operatorname{arccosech}(\coth 2x)$
    (iii) $\operatorname{arcsech}(\tanh x)$
    (iv) $\operatorname{arctanh}(\sinh x)$
    (v) $\operatorname{arccoth}(3x^2 - 1)$.

6

Parametric Equations

Certain cartesian functions may be difficult to sketch. If x and y, however, are expressed in terms of a third variable, t, called the parameter, the functions may be sketched easier. Consider the cartesian function

$$y^2 = x^2(x-1) \qquad \ldots (1)$$

whose parametric equations are $x = t^2 + 1$ and $y = t(t^2 + 1)$.

The left hand side of (1)

$$y^2 = t^2(t^2+1)^2 = t^2(t^4 + 2t^2 + 1)$$

the right hand side of (1)

$$x^2(x-1) = (t^2+1)^2(t^2+1-1) = t^2(t^2+1)^2.$$

Therefore the parametric equations of $y^2 = x^2(x-1)$ are $x = t^2 + 1$ and $y = t(t^2+1)$.

Sketch the cartesian equation $y^2 = x^2(x-1)$
if $x = 0$, $y = 0$; if $x \geq 1$, y exists
if $x = 1$, $y = 0$; if $x = 2$, $y = \pm 2$ and
if $x = 3$, $y^2 = 9(2)$, $y = \pm 3\sqrt{2}$ and so on.

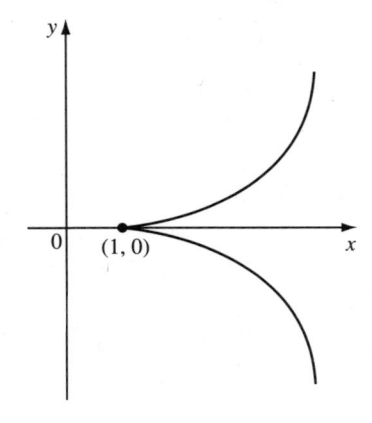

Fig. 4-I/15

The curve is symmetrical about the x-axis.
Sketch the parametric equations

$$x = t^2 + 1, y = t(t^2 + 1).$$

if $t = 0$;	$x = 1$,	$y = 0$
if $t = 1$;	$x = 2$,	$y = 2$
if $t = 2$;	$x = 5$,	$y = 10$
if $t = -1$;	$x = 2$,	$y = -2$
if $t = -2$;	$x = 5$,	$y = -10$

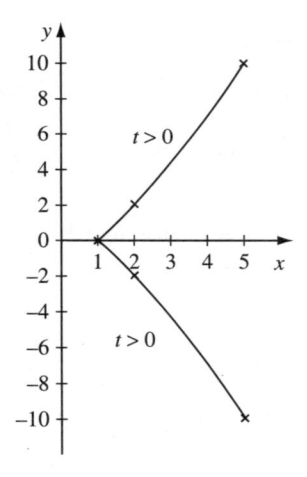

Fig. 4-I/16

Differentiating $x = t^2 + 1$ and $y = t(t^2 + 1) = t^3 + t$ with respect to t, we have

$$\frac{dx}{dt} = 2t \text{ and } \frac{dy}{dt} = 3t^2 + 1$$

$$\frac{\frac{dy}{dt}}{\frac{dx}{dt}} = \frac{dy}{dx} = \frac{3t^2 + 1}{2t}.$$

If t is positive, $\frac{dy}{dx}$ is positive. If t is negative, $\frac{dy}{dx}$ is negative.

33

WORKED EXAMPLE 26

Determine the cartesian equations of the following parametric equations:-

(i) $x = a \cos \Theta$
$y = b \sin \Theta$

(ii) $x = at^2$
$y = 2at$

(iii) $x = ct$
$y = \dfrac{c}{t}$

(iv) $x = r \cos \Theta$
$y = r \sin \Theta$

(v) $x = a \cosh \Theta$
$y = b \sinh \Theta$.

Solution 26

(i) $x = a \cos \Theta$, $\cos \Theta = \dfrac{x}{a}$ ...(1)

$y = b \sin \Theta$, $\sin \Theta = \dfrac{y}{b}$ ...(2)

squaring up both sides of (1) and (2) and adding

$$\cos^2 \Theta + \sin^2 \Theta = \left(\dfrac{x}{a}\right)^2 + \left(\dfrac{y}{b}\right)^2 = 1$$

$$\boxed{\dfrac{x^2}{a^2} + \dfrac{y^2}{b^2} = 1, \text{ an ellipse.}}$$

(ii) $x = at^2$ ...(1)
$y = 2at$ ...(2)

From (2) $t = \dfrac{y}{2a}$, substitute in (1)

$$x = a\left(\dfrac{y}{2a}\right)^2 = \dfrac{y^2}{4a}$$

$$\boxed{y^2 = 4ax, \text{ a parabola.}}$$

(iii) $x = ct$ ...(1)
$y = \dfrac{c}{t}$ ...(2)

Multiplying (1) and (2) $xy = ct\left(\dfrac{c}{t}\right) = c^2$

$$\boxed{xy = c^2, \text{ a rectangular hyperbola}}$$

(iv) $x = r \cos \Theta$ ...(1) $\cos \Theta = \dfrac{x}{r}$ ...(3)

$y = r \sin \Theta$ ...(2) $\sin \Theta = \dfrac{y}{r}$ ...(4)

squaring up both sides of (3) and (4) and adding:-

$$\cos^2 \Theta + \sin^2 \Theta = \dfrac{x^2}{r^2} + \dfrac{y^2}{r^2} = 1$$

$$\boxed{x^2 + y^2 = r^2, \text{ a circle}}$$

(v) $x = a \cosh \Theta$, ...(1) $\cosh \Theta = \dfrac{x}{a}$ ...(3)

$y = b \sinh \Theta$, ...(2) $\sinh \Theta = \dfrac{y}{b}$ ...(4)

squaring up both sides of (3) and (4) and subtracting

$$\boxed{\cosh^2 \Theta - \sinh^2 \Theta = \dfrac{x^2}{a^2} - \dfrac{y^2}{b^2} = 1, \text{ a hyperbola}}$$

Determine $\dfrac{dy}{dx}$ and $\dfrac{d^2y}{dx^2}$ for the parametric equations:-

(i) $x = a \cos \Theta$ $\quad \dfrac{dx}{d\Theta} = -a \sin \Theta$

$y = b \sin \Theta$ $\quad \dfrac{dy}{d\Theta} = b \cos \Theta$

$$\dfrac{dy}{dx} = \dfrac{\frac{dy}{d\Theta}}{\frac{dx}{d\Theta}} = -\dfrac{b}{a} \cot \Theta$$

(ii) $x = at^2$, $\quad \dfrac{dx}{dt} = 2at$

$y = 2at$ $\quad \dfrac{dy}{dt} = 2a$

$$\dfrac{dy}{dx} = \dfrac{\frac{dy}{dt}}{\frac{dx}{dt}} = \dfrac{2a}{2at} = \dfrac{1}{t}$$

(iii) $x = ct$ $\quad \dfrac{dx}{dt} = c$

$y = \dfrac{c}{t} = ct^{-1}$, $\quad \dfrac{dy}{dt} = -\dfrac{c}{t^2}$

$$\dfrac{dy}{dx} = \dfrac{\frac{dy}{dt}}{\frac{dx}{dt}} = \dfrac{-\frac{c^2}{t}}{c} = -\dfrac{1}{t^2}$$

(iv) $x = r \cos \Theta$ $\quad \dfrac{dx}{d\Theta} = -r \sin \Theta$

$y = r \sin \Theta$ $\quad \dfrac{dy}{d\Theta} = r \cos \Theta$

$$\frac{dy}{dx} = \frac{\frac{dy}{d\Theta}}{\frac{dx}{d\Theta}}$$

$$= \frac{r\cos\Theta}{-r\sin\Theta} = -\cot\Theta$$

(v) $x = a\cosh\Theta$ $\quad \frac{dx}{d\Theta} = a\sinh\Theta$

$y = b\sinh\Theta$ $\quad \frac{dy}{d\Theta} = b\cosh\Theta$

$$\frac{dy}{dx} = \frac{b\cosh\Theta}{a\sinh\Theta} = \frac{b}{a}\coth\Theta.$$

(i) $\displaystyle \frac{d^2y}{dx^2} = -\frac{b}{a}(-\mathrm{cosec}^2\,\Theta)\frac{d\Theta}{dx}$

$\displaystyle = \frac{b}{a}\mathrm{cosec}^2\,\Theta\left(\frac{1}{-a\sin\Theta}\right)$

$\displaystyle = -\frac{b}{a^2}\mathrm{cosec}^3\,\Theta$

(ii) $\displaystyle \frac{d^2y}{dx^2} = -\frac{1}{t^2}\cdot\frac{dt}{dx} = -\frac{1}{t^2}\cdot\frac{1}{2at} = -\frac{1}{2at^3}$

(iii) $\displaystyle \frac{d^2y}{dx^2} = \frac{2}{t^3}\cdot\frac{dt}{dx} = \frac{2}{t^3}\cdot\frac{1}{c} = \frac{2}{ct^3}$

(iv) $\displaystyle \frac{d^2y}{dx^2} = -(-\mathrm{cosec}^2\,\Theta)\frac{d\Theta}{dx}$

$\displaystyle = \mathrm{cosec}^2\,\Theta\,\frac{1}{-r\sin\Theta} = -\frac{1}{r}\mathrm{cosec}^3\,\Theta$

(v) $\displaystyle \frac{d^2y}{dx^2} = -\frac{b}{a}\mathrm{cosech}^2\,\Theta\,\frac{d\Theta}{dx}$

$\displaystyle = -\frac{b}{a}\mathrm{cosech}^2\,\Theta\,\frac{1}{a\sinh\Theta}$

$\displaystyle = -\frac{b}{a^2}\mathrm{cosech}^3\,\Theta.$

Exercises 6

1. Obtain $\displaystyle \frac{dy}{dx}$ in terms of the parameter t, if $x = 2\sinh t$, $y = 3\cosh t$.

2. (a) If $x = t - \sin t$ and $y = 1 - \cos t$, find
   (i) $\displaystyle \frac{dy}{dx}$ (ii) $\displaystyle \frac{d^2y}{dx^2}$ in terms of half angles.

   (b) Sketch the curve given by the parametric equations in (a) for $0 \le t \le 2\pi$.

3. If $x = ct$ and $y = \displaystyle \frac{c}{t}$, find
   (i) $\displaystyle \frac{dy}{dt}$
   (ii) $\displaystyle \frac{d^2y}{dx^2}$ in terms of t. Sketch the curve.

4. Sketch the curve given parametrically by $x = 2t^2$ and $y = 2t^3$.

5. If $x = 1 + t^2$ and $y = 2t - 1$, find
   (i) $\displaystyle \frac{dy}{dx}$
   (ii) $\displaystyle \frac{d^2y}{dx^2}$ in terms of t. Sketch the curve.

6. Sketch the curve given parametrically by $x = 4t^2$, $y = 4t$. You may consider the gradient $\displaystyle \frac{dy}{dx}$.

7. If $x = 2\sin t$ and $y = 2\cos^3 t$, determine
   (i) $\displaystyle \frac{dy}{dx}$ and
   (ii) $\displaystyle \frac{d^2y}{dx^2}$ and hence determine the points of inflexion. Sketch the curve.

8. A curve is given by the parametric equations $x = 2\cos t + (t+3)\sin t$, $y = 2\sin t - (t+3)\cos t$, $0 \le t \le 2\pi$, $t \ne -3$.
   Determine $\displaystyle \frac{dy}{dx}$.

9. The parametric equations of a curve are $x = 2\cos\Theta$ and $y = 4\sin\Theta$.
   Find $\displaystyle \frac{dy}{dx}$.

10. A curve has parametric equations $x = t + e^t$, $y = 2t - e^{2t}$.
    Determine $\displaystyle \frac{dx}{dt}, \frac{dy}{dt}$ and hence $\displaystyle \frac{dy}{dx}$ and $\displaystyle \frac{d^2y}{dx^2}$.

11. A curve has parametric equations $x = 3t$ and $y = 3\ln\sec t$.
    Determine the derivatives:-
    (i) $\displaystyle \frac{dy}{dx}$ and (ii) $\displaystyle \frac{d^2y}{dx^2}$.

12. A curve has parametric equations of $x = t - \sin t$, $y = 1 - \cos t$. Find $\dfrac{dy}{dx}$.

13. A curve has parametric equations $x = 3\cos t$ and $y = \cos 2t$.

    Determine (i) $\dfrac{dy}{dx}$ and (ii) $\dfrac{d^2y}{dx^2}$.

14. A curve is given by the equations
    $x = 3\sin 2t(1 - \cos 2t)$,
    $y = 3\cos 2t(1 + \cos 2t)$. Find $\dfrac{dy}{dx}$.

15. A curve is defined by the parametric equations
    $x = 2\cos\Theta - \cos 2\Theta$, $y = 2\sin\Theta - \sin 2\Theta$.
    Determine $\dfrac{dy}{dx}$ and $\dfrac{d^2y}{dx^2}$.

16. Find $\dfrac{dy}{dx}$ at $t = \dfrac{\pi}{2}$ for a curve whose parametric equations are
    $x = t\sin t - \cos t - 1$, $y = \sin t - t\cos t$.

17. A curve is defined by the parametric equations
    $x = 3(\Theta - \sin\Theta)$, $y = 3(1 - \cos\Theta)$.
    Show that $\dfrac{d^2y}{dx^2} = -\dfrac{1}{12}\operatorname{cosec}^4\dfrac{\Theta}{2}$.

18. A curve is given parametrically by the equations:-
    $x = t - \tanh t$, $y = \operatorname{sech} t$. Find $\dfrac{dy}{dx}$ and $\dfrac{d^2y}{dx^2}$.
    Hence determine the value of $\left(\dfrac{dy}{dx}\right)^3 - \left(\dfrac{d^2y}{dx^2}\right) + \dfrac{y}{x}$.

7

Applications of Differentiation

Second and Higher Derivatives of a Function

Notation of Second Derivative

$\frac{d}{dx}\left(\frac{dy}{dx}\right)$ = the rate of change of $\frac{dy}{dx}$ with respect to x

$= \frac{d^2y}{dx^2}$

$\frac{dy}{dx} = f'(x)$ when $y = f(x)$

$\frac{d^2y}{dx^2} = f''(x)$ = the second derivative.

The Meaning of the Second Derivative

The first derivative or the primitive of a function denotes the gradient of a function.

$\frac{dy}{dx}$ may be positive, $\frac{dy}{dx} > 0$

$\frac{dy}{dx}$ may be negative, $\frac{dy}{dx} < 0$

$\frac{dy}{dx}$ may be zero, $\frac{dy}{dx} = 0$.

To illustrate the various cases above, consider the parabola with a maximum and a minimum. $\frac{dy}{dx}$ is changing from a positive to a negative, the rate of change of $\frac{dy}{dx}$ is $\frac{d^2y}{dx^2} < 0$,

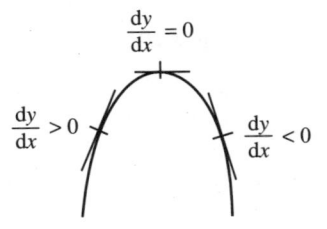

Fig. 4-I/17

negative, $\frac{d}{dx}\left(\frac{dy}{dx}\right) = \frac{d^2y}{dx^2} < 0$. Therefore, for a maximum the rate of change of the gradient is negative. At the peak, it is neither positive or negative, it is $\frac{dy}{dx} = 0$ for a maximum. Consider a minimum.

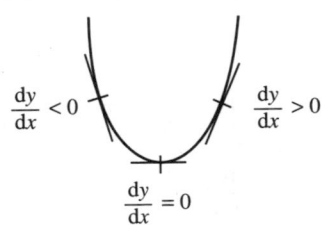

Fig. 4-I/18

The rate of change of $\frac{dy}{dx}$ is positive, $\frac{dy}{dx}$ changes from negative to positive, via a zero gradient.

WORKED EXAMPLE 27

Sketch the graph $y = \sin x$ and determine the gradients at $x = 0, \frac{\pi}{4}, \frac{\pi}{2}, \frac{3\pi}{4}, \pi, \frac{5\pi}{4}, \frac{3\pi}{2}, \frac{7\pi}{4}$ and 2π.

Illustrate the primitive or first derivative and second derivatives for the maximum and minimum values of the function.

37

Solution 27

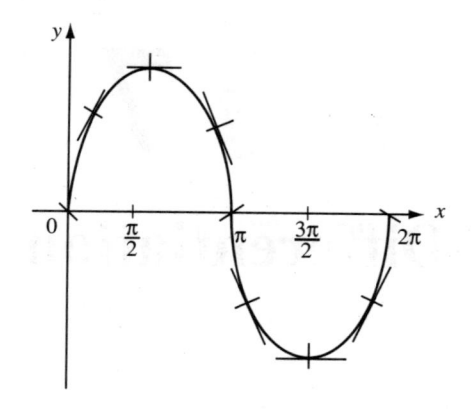

Fig. 4-I/19

$y = \sin x$ $\qquad \dfrac{dy}{dx} = \cos x;$

at $x = 0,$ $\qquad \dfrac{dy}{dx} = \cos 0 = 1;$

at $x = \dfrac{\pi}{4},$ $\qquad \dfrac{dy}{dx} = \cos \dfrac{\pi}{4} = 0.707;$

at $x = \dfrac{\pi}{2},$ $\qquad \dfrac{dy}{dx} = \cos \dfrac{\pi}{2} = 0;$

at $x = \dfrac{3\pi}{4},$ $\qquad \dfrac{dy}{dx} = \cos \dfrac{3\pi}{4} = -0.707;$

at $x = \pi,$ $\qquad \dfrac{dy}{dx} = \cos \pi = -1;$

at $x = \dfrac{5\pi}{4},$ $\qquad \dfrac{dy}{dx} = \cos \dfrac{5\pi}{4} = -0.707;$

at $x = \dfrac{3\pi}{2},$ $\qquad \dfrac{dy}{dx} = 0;$

at $x = \dfrac{7\pi}{4},$ $\qquad \dfrac{dy}{dx} = \cos \dfrac{7\pi}{4} = 0.707;$ and

at $x = 2\pi,$ $\qquad \dfrac{dy}{dx} = 1.$

For a maximum

at $x = \dfrac{\pi}{4}, \dfrac{dy}{dx} > 0 \qquad$ at $x = \dfrac{\pi}{2}, \dfrac{dy}{dx} = 0,$

at $x = \dfrac{3\pi}{4}, \dfrac{dy}{dx} < 0$

$$\boxed{\dfrac{dy}{dx}\left(\dfrac{dy}{dx}\right) < 0 \text{ or } \dfrac{d^2y}{dx^2} < 0.}$$

For a minimum

at $x = \dfrac{5\pi}{4}, \dfrac{dy}{dx} < 0 \qquad$ at $x = \dfrac{3\pi}{2}, \dfrac{dy}{dx} = \cos \dfrac{3\pi}{2} = 0$

at $x = \dfrac{7\pi}{4}, \dfrac{dy}{dx} > 0 \qquad \boxed{\dfrac{dy}{dx}\left(\dfrac{dy}{dx}\right) > 0 \text{ or } \dfrac{d^2y}{dx^2} > 0.}$

Exercises 7

1. Determine the first and second derivatives for the following functions:-

   (i) $y = 3x^2 - 5x + 7$ (ii) $x = t - 6t^2 + 7t^3$

   (iii) $u = 3v^2 + 5v - 1$ (iv) $w = 3z^2 - z - 4.$

2. Determine the second derivatives of the following functions and simplify:-

   (i) $y = \dfrac{3x^2 - 1}{x + 1}$

   (ii) $y = e^x + \sin x$

   (iii) $y = \dfrac{e^{-x}}{\cos 2x}$

   (iv) $y = 3 \sin 2x - 5 \cos 2x$

   (v) $y = \sin^2 x.$

3. A body is moving along a straight line and its distance x metres from a fixed point on the line after a time t seconds is given by $x = 2t^3 - 3t^2 + 4t + 5.$

   Find

   (i) the velocity of the body after 1 s,

   (ii) the velocity of the body at $t = 0$ s,

   (iii) the velocity of the body after 5 s from the start,

   (iv) the acceleration at the start and after 2 s,

   (v) the displacements after 2, 3, and 5 s.

4. If the distance s metres a body moves after t s is given by $s = 30t^2 - 3t + 5,$

   find

   (i) its velocity after 3 s,

   (ii) its acceleration,

   (iii) the distance the body has travelled before coming to rest,

   (iv) the time when the velocity is 57 m/s,

   (v) the velocity after 10 s.

5. A body is falling freely from rest under gravity (= 9.81 m/s^2); the distance s metres travelled is given by the expression $s = 20t^2$, where t is the time in seconds.

    Find

    (i) the velocity after t seconds,

    (ii) the velocity after 1 second,

    (iii) the time taken for the body to fall 1500 m,

    (iv) the acceleration.

6. Find $\dfrac{d^2y}{dx^2}$ for the following functions:-

    (i) $y = 3 \sin 2x - 5 \cos 2x$

    (ii) $y = 3x^3 - 2x^2 + x - 1$

    (iii) $y = 4e^{-2x} - 5e^{3x}$

    (iv) $y = e^{3x} - \cos 3x + \sin 3x$

    (v) $y = 5 \ln x + x \sin 2x$.

7. Determine the first and second derivatives of the following functions:-

    (i) $x = 5t^5 - 4t^4 + 3t^3 - 2t^2 + t - 1$

    (ii) $x = \sin t - \cos t$

    (iii) $x = e^t \sin 2t$

    (iv) $x = \dfrac{e^{2t}}{(1+t)}$

    (v) $x = e^{\sin t}$.

8. Define velocity and acceleration and determine the velocity and acceleration for the functions in exercise 7 at $t = 0$.

9. Determine the second derivatives for the following functions:-

    (i) $y = \dfrac{x^2 + 1}{x - 1}$

    (ii) $y = x^2 \sin x$

    (iii) $y = \dfrac{\cos x}{e^{3x}}$

    (iv) $y = \dfrac{\ln x}{(1+x)^2}$

    (v) $y = \dfrac{e^x}{(1+x)}$.

10. Evaluate the second derivatives for the following functions at $x = \dfrac{\pi}{2}$.

    (i) $y = \sin x \cos 2x$

    (ii) $y = e^x \tan \dfrac{x}{2}$

    (iii) $y = x^2 \ln x$

    (iv) $y = 5 \cos 3x - 4 \sin 4x$

11. Determine $\dfrac{d^2y}{dx^2}$ for $y = 3 \cos 5kx$, where k is a constant, and at $x = \dfrac{\pi}{k}$.

12. What is the significance of $\dfrac{dy}{dx}$ and $\dfrac{d^2y}{dx^2}$ of a function? Illustrate your answers clearly with the aid of sketches.

13. Find the second derivative for the function

    $y = e^x \sin 2x$.

14. Find the second derivative of the function

    $y = e^{-2x} \cos x$.

15. A particle moves s m in time t seconds given by the relation $s = 3t^3 - t^2 + t + 7$. Find the velocity and the acceleration of the particle after 5 seconds.

16. Find the $\dfrac{d^2y}{dx^2}$ of the following quotients:-

    (i) $\dfrac{1 + \cos \Theta}{\sin \Theta}$

    (ii) $\dfrac{\sin x}{e^{2x}}$

    (iii) $\cot \Theta$.

17. If $i = I_m \sin\left(2\pi ft - \dfrac{\pi}{3}\right)$, determine $\dfrac{d^2i}{dt^2}$.

18. If $v = V_m \cos\left(2\pi ft + \dfrac{\pi}{6}\right)$, determine $\dfrac{d^2v}{dt^2}$.

19. The volume of a sphere is given as $V = \dfrac{4}{3}\pi R^3$; find $\dfrac{d^2V}{dR^2}$.

20. The volume of a cone is given as $V = \dfrac{1}{3}\pi r^2 h$ and $h = 3r$; find $\dfrac{d^2V}{dr^2}$.

8
Tangents and Normals

WORKED EXAMPLE 28

Determine the equation of the tangent at a point (1, 2) of the curve $y = 2x^2$.

Solution 28

The equation of a straight line is given by $y = mx + c$ where m is the gradient, and c is the intercept.

The gradient of the curve is found by differentiating $y = 2x^2$

$$\frac{dy}{dx} = 4x.$$

At $x = 1$, $\frac{dy}{dx} = 4$, the equation of the tangent is $y = 4x + c$, this line passes through the point (1, 2), $x = 1$ and $y = 2$;

$$2 = 4 \times 1 + c$$
$$c = -2$$

therefore the equation of the line is

$$\boxed{y = 4x - 2}$$

To Find the Angle between Two Lines

Consider two lines which intersect at an angle Θ. Let α and β be the angles that the lines make with the horizontal axis.

$$\tan \alpha = m_1, \text{ and } \tan \beta = m_2$$

$\Theta = \alpha - \beta$, taking tangents on both sides, we have

$$\tan \Theta = \frac{\tan \alpha - \tan \beta}{1 + \tan \alpha \tan \beta} = \frac{m_1 - m_2}{1 + m_1 m_2}.$$

The lines are parallel when $m_1 = m_2 \quad \tan \Theta = 0$.

The lines are perpendicular when $\Theta = 90°$, $\tan 90° = \infty$, that is, $1 + m_1 m_2 = 0$.

The gradient of a normal is perpendicular to the tangent at the same point.

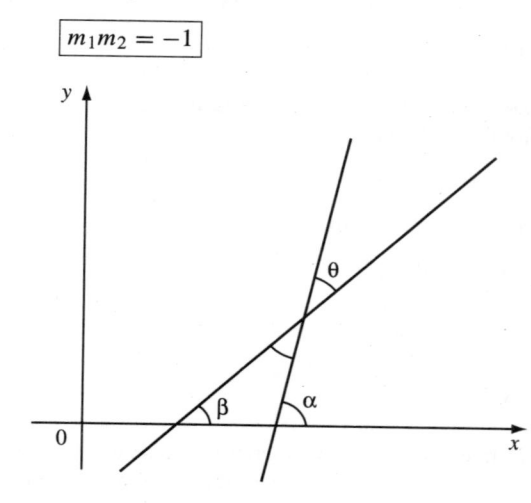

$$\boxed{m_1 m_2 = -1}$$

Fig. 4-I/20

A curve is given by the parametric equations.

$$x = \Theta - \sin \Theta \quad \ldots (1)$$
$$y = 1 - \cos \Theta \quad \ldots (2)$$

Sketch the curve $0 \leq \Theta \leq 2\pi$.

Θ^c	0	$\frac{\pi}{6}$	$\frac{\pi}{3}$	$\frac{\pi}{2}$	$\frac{2\pi}{3}$	$\frac{5\pi}{6}$
$\sin \Theta^c$	0	0.5	0.866	1	0.866	0.5
$\cos \Theta^c$	1	0.866	0.5	0	-0.5	-0.866
$y = 1 - \cos \Theta$	0	0.134	0.5	1	1.5	1.866
$x = \Theta - \sin \Theta$	0	0.024	0.181	0.571	1.228	2.118

cont...

π	$\dfrac{7\pi}{6}$	$\dfrac{4\pi}{3}$	$\dfrac{3\pi}{2}$	$\dfrac{5\pi}{3}$	$\dfrac{11\pi}{6}$	2π
0	−0.5	−0.866	−1	−0.866	−0.5	0
−1	−0.866	−0.5	0	0.5	0.866	1
2	1.866	1.5	1	0.5	0.134	0
3.142	4.165	5.055	5.712	6.102	6.260	6.283

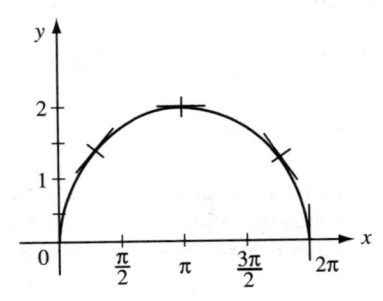

Fig. 4-I/21

Determine $\dfrac{dy}{dx}$.

Differentiating (1) and (2)

$$\dfrac{dx}{d\Theta} = 1 - \cos\Theta \qquad \dfrac{dy}{d\Theta} = \sin\Theta$$

$$\dfrac{dy}{dx} = \dfrac{\frac{dy}{d\Theta}}{\frac{dx}{d\Theta}} = \dfrac{\sin\Theta}{1-\cos\Theta} = \dfrac{2\sin\frac{\Theta}{2}\cos\frac{\Theta}{2}}{1-\left(2\cos^2\frac{\Theta}{2}-1\right)}$$

$$\dfrac{dy}{dx} = \dfrac{2\sin\frac{\Theta}{2}\cos\frac{\Theta}{2}}{2\left(1-\cos^2\frac{\Theta}{2}\right)} = \dfrac{2\sin\frac{\Theta}{2}\cos\frac{\Theta}{2}}{2\sin^2\frac{\Theta}{2}} = \cot\dfrac{\Theta}{2}$$

Determine the values of $\dfrac{dy}{dx}$ at Θ equal to the following:

(i) $\dfrac{\pi}{2}$

(ii) π

(iii) $\dfrac{3\pi}{2}$

(iv) 2π.

(i) $\dfrac{dy}{dx} = \cot\dfrac{\Theta}{2} = \cot\dfrac{\pi}{4} = 1$

(ii) $\dfrac{dy}{dx} = \cot\dfrac{\Theta}{2} = \cot\dfrac{\pi}{2} = 0$

(iii) $\dfrac{dy}{dx} = \cot\dfrac{\Theta}{2} = \cot\dfrac{3\pi}{4} = -1$

(iv) $\dfrac{dy}{dx} = \cot\dfrac{\Theta}{2} = \cot\pi = -\infty$.

Determine the equations of the tangents and normals at $\Theta = \dfrac{\pi}{2}$ and $\Theta = \dfrac{3\pi}{2}$.

At $\Theta = \dfrac{\pi}{2}$ $\quad x = \Theta - \sin\Theta = \dfrac{\pi}{2} - 1$

$y = 1 - \cos\Theta = 1$

gradient $= 1$,

$y = mx + c = 1\cdot x + c$

$1 = \dfrac{\pi}{2} - 1 + c, c = 2 - \dfrac{\pi}{2}$

Therefore the equation of the tangent is

$$\boxed{y = x + 2 - \dfrac{\pi}{2}}$$

the gradient of the normal is -1,

$y = -x + c, \quad 1 = -\left(\dfrac{\pi}{2} - 1\right) + c \quad c = \dfrac{\pi}{2}$

Therefore the equation of the normal is $\boxed{y = -x + \dfrac{\pi}{2}}$

At $\Theta = \dfrac{3\pi}{2}, x = \Theta - \sin\Theta = \dfrac{3\pi}{2} + 1$,

$y = 1 - \cos\Theta = 1$ and gradient $= -1$

$y = -x + c \quad 1 = -\left(\dfrac{3\pi}{2} + 1\right) + c \quad c = 2 + \dfrac{3\pi}{2}$

Therefore the equation of the tangent is

$$\boxed{y = -x + 2 + \dfrac{3\pi}{2}}$$ the gradient of the normal is 1

$y = x + c \quad 1 = \dfrac{3\pi}{2} + 1 + c \quad c = -\dfrac{3\pi}{2}$

Therefore the equation of the normal is $\boxed{y = x - \dfrac{3\pi}{2}}$

The cartesian equation of the cycloid.

Eliminate the parameter between (1) and (2).

$x = \Theta - \sin\Theta \qquad \sin\Theta = \Theta - x$
$y = 1 - \cos\Theta \qquad \cos\Theta = 1 - y$
$\sin^2\Theta + \cos^2\Theta = 1 \qquad (\Theta - x)^2 + (1 - y)^2 = 1$
$\Theta - x = \pm\sqrt{1 - (1-y)^2} = \pm\sqrt{(2-y)y}$
$\Theta = x \pm \sqrt{y(2-y)}$
$1 - y = \cos\left[x \pm \sqrt{y(2-y)}\right]$
$0 \le y \le 2$ and $0 \le x \le 2\pi$.

The cartesian equation is rather complicated.

WORKED EXAMPLE 29

Determine the equation of the normal at the point (1, 2) of the curve $y = 2x^2$.

Solution 29

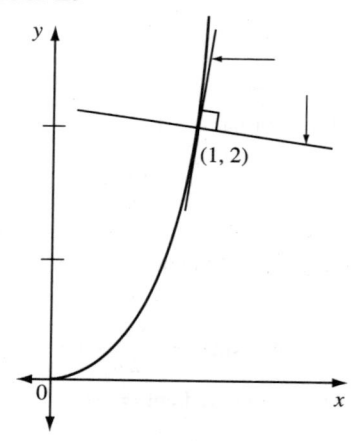

Fig. 4-I/22

The equation of the normal is given by the equation $y = mx + c$. If m_1 is the gradient of the tangent, m_2 is the gradient of the normal, since the tangent and normal are at $90°$, $m_1 m_2 = -1$; the gradient of the normal, $m_2 = -\dfrac{1}{m_1} = -\dfrac{1}{4}$.

The equation of the normal is $y = -\dfrac{1}{4}x + c$; this equation passes through the point (1, 2), $2 = -\dfrac{1}{4} + c$, $c = \dfrac{9}{4}$, then $y = -\dfrac{1}{4}x + \dfrac{9}{4}$ $\boxed{4y + x = 9}$ is the equation of the normal.

WORKED EXAMPLE 30

Determine the equations of the tangent and normal to the given curve at the given point adjacent to each curve.
c
(i) $y = 3x^2 + 5$ when $x = -1$
(ii) $y = 3x^2 - 5x - 1$ when $x = 2$.

Solution 30

(i) $y = 3x^2 + 5$, $\dfrac{dy}{dx} = 6x$ at any point,

when $x = -1$, $\dfrac{dy}{dx} = -6$.

The equation of the tangent is $y = -6x + c$, it passes through the point where $x = -1$ and $y = 3(-1)^2 + 5 = 8$, $8 = -6(-1) + c$, $c = 8 - 6 = 2$

$\boxed{y = -6x + 2}$

The gradient of the normal,

$m_2 = -\dfrac{1}{m_1} = -\dfrac{1}{-6} = \dfrac{1}{6}$; the equation of the normal is $y = \dfrac{1}{6}x + c$, it passes through $(-1, 8)$,

$8 = -\dfrac{1}{6} + c$, $c = 8 + \dfrac{1}{6} = \dfrac{49}{6}$, $y = \dfrac{1}{6}x + \dfrac{49}{6}$

$\boxed{6y = x + 49}$

(ii) $y = 3x^2 - 5x - 1$, $\dfrac{dy}{dx} = 6x - 5$, the gradient at any point, when $x = 2$, $\dfrac{dy}{dx} = 6 \times 2 - 5 = 7$.

The equation of the tangent is $y = 7x + c$, it passes through the point $x = 2$

$y = 3(2)^2 - 5(2) - 1 = 12 - 10 - 1 = 1$,

$1 = 7 \times 2 + c$

$c = 1 - 14 = -13$

$\boxed{y = 7x - 13}$

The gradient of the normal is found from $m_1 m_2 = -1$, $m_2 = -\dfrac{1}{m_1} = -\dfrac{1}{7}$

$$y = -\frac{1}{7}x + c$$

this passes through the point (2, 1)

$$1 = -\frac{2}{7} + c$$

$$c = \frac{9}{7}$$

equation of normal is $\quad y = -\frac{1}{7}x + \frac{9}{7}$

$$\boxed{7y + x = 9}$$

WORKED EXAMPLE 31

Determine the equations of the tangent and normal to the curve $y = -x^2 - 5x + 6$ at the points when it cuts the x-axis and y-axis.

Solution 31

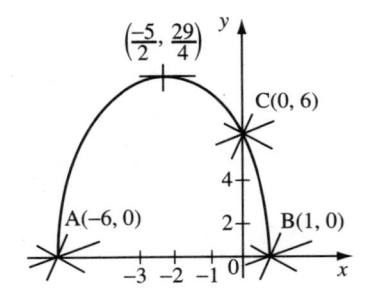

Fig. 4-I/23

when $x = 0$, $y = 6$; the curve has a maximum at

$$x = -\frac{b}{2a} = \frac{-(-5)}{-2} = -\frac{5}{2},$$

$$y_{max} = -\left(-\frac{5}{2}\right)^2 - 5\left(-\frac{5}{2}\right) + 6, \left(-\frac{5}{2}, \frac{49}{4}\right); \text{ when }$$
$y = 0$, $x = 1$ and $x = -6$.

The equations of the tangents and normals at $(-6, 0)$, $(1, 0)$ and $(0, 6)$ are required.

At $A(-6, 0)$

The equation of the tangent

$$y = mx + c, \qquad y = \frac{dy}{dx}x + c,$$

$$y = -x^2 - 5x + 6, \quad \frac{dy}{dx} = -2x - 5$$

when $x = -6$, $\frac{dy}{dx} = -2(-6) - 5 = 7$

$$y = 7x + c$$

when $x = -6$, $y = 0$, $c = 42$, $\quad \boxed{y = 7x + 42}$

The gradient of the normal is $-\frac{1}{7}$ $\quad y = -\frac{1}{7}x + c$

when $x = -6$, $y = 0$, $0 = \frac{-(-6)}{7} + c$, $c = -\frac{6}{7}$

$$y = -\frac{1}{7}x - \frac{6}{7}$$

the equation of the normal

$$\boxed{7y + x + 6 = 0}$$

At $B(1, 0)$

The equation of the tangent.

$$y = -x^2 - 5x + 6, \quad \frac{dy}{dx} = -2x - 5, \text{ when } x = 1,$$

$$\frac{dy}{dx} = -7$$

$y = -7x + c$, when $x = 1$, $y = 0$, $c = 7$

$$\boxed{y = -7x + 7}$$

The gradient of the normal is $\frac{1}{7}$ $\quad y = \frac{1}{7}x + c$

when $x = 1$, $y = 0$, $\quad c = -\frac{1}{7}, y = \frac{1}{7}x - \frac{1}{7}$

$$\boxed{7y = x - 1}$$

At $C(0, 6)$

The equation of the tangent.

When $x = 0$, $\frac{dy}{dx} = -5$, $y = -5x + c$, when $x = 0$, $y = 6$, $c = 6$,

$$\boxed{y = -5x + 6}$$

The equation of the normal

$m = \frac{1}{5}$, $y = \frac{1}{5}x + c$, when $x = 0$, $y = 6$, $c = 6$

$$\boxed{5y = x + 30}$$

Exercises 8

1. The parametric equations of a curve are $x = a(2\Theta - \sin 2\Theta)$ and $y = a(1 - \cos 2\Theta)$.
   Determine $\dfrac{dy}{dx}$ and hence find the equations of the tangent and normal to the curve at the point P where $\Theta = \dfrac{\pi}{4}$.

2. Determine the equations of the tangent and normal at any point t on the curve having parametric equations $x = 2\cos t - \cos 2t$, $y = 2\sin t - \sin 2t$.

3. Find the equation of the normal at a general point Θ on the ellipse $x = 2\cos\Theta$ and $y = 3\sin\Theta$.

4. Find the equations of the normal and tangent at the point where $x = \dfrac{5}{8}$ of the curve whose parametric equations are given by $x = 5\sin^3\Theta$, $y = 5\cos^3\Theta$.

5. Find the equations of the tangents to the curve $x^2 + y^2 - x - y - 2 = 0$ at the points $(1, -1)$ and $(1, 2)$.

6. Find the equation of the normal at the point $(1, -2)$ of the curve $y^2 = 4x$.

7. Find the equations of the tangent and normal at the point of $(-1, -9)$ of the curve $xy = 9$.

9

Small Increments and Approximations L'Hôpital's Rule Rates of Change

Small Increments

If $y = f(x)$ and if δx, δy are respectively the increment in x and the corresponding increment in y, the limiting value of the ratio $\dfrac{\delta y}{\delta x}$ when δx approaches zero is, by definition, $\dfrac{dy}{dx}$ or $f'(x)$.

$$\frac{\delta y}{\delta x} \to \frac{dy}{dx} = f'(x) \text{ as } \delta x \to 0$$

WORKED EXAMPLE 32

The area of a triangle is given as

$$A = \frac{1}{2}ab\sin\Theta$$

when Θ is the angle between the two sides a, b where $a = 25$ cm, $b = 45$ cm and $\Theta = 45°$. If Θ changes then A changes

$$\frac{dA}{d\Theta} = \frac{1}{2}ab\cos\Theta$$

For an error in Θ of $1'$, determine the corresponding error in the area.

Solution 32

$$\frac{\delta A}{\delta\Theta} \approx \frac{dA}{d\Theta} = \frac{1}{2}ab\cos\Theta$$

$$\delta A = \frac{1}{2}ab\cos\Theta\,\delta\Theta.$$

If $a = 25$ cm, $b = 45$ cm, $\Theta = 45°$

$$\delta A = \frac{1}{2} \times 25 \times 45 \times \cos 45° \,\delta\Theta$$

$\delta\Theta = 1'$

$$= \left(\frac{1}{60}\right)^\circ = \left(\frac{1}{60} \times \frac{\pi}{180}\right)^c = 2.9088821 \times 10^{-4}$$

$$\delta A = \frac{1}{2} \times 25 \times 45 \times \cos 45° \times 2.9088821 \times 10^{-4}$$

$$= 0.11568 \text{ cm}^2.$$

WORKED EXAMPLE 33

If $y^5 = x^5(x - 5)$, find using calculus the increase in y when x increases from 3.563 to 3.564, giving your answer correct to five significant figures. State the new value of y in five significant figures.

Solution 33

$y^5 = x^5(x - 5)$, differentiating with respect to x,

$5y^4\dfrac{dy}{dx} = 5x^4(x - 5) + x^5$ which can be written as

$$5y^4\delta y = \left(5x^5 - 25x^4 + x^5\right)\delta x \text{ or}$$

$$5y^4\delta y = \left(6x^5 - 25x^4\right)\delta x$$

$$\delta y = \frac{6x^5 - 25x^4}{5y^4}\delta x$$

$$= \frac{6(3.563)^5 - 25(3.563)^4}{5y^4}(0.001)$$

where $y^5 = (3.563)^5(3.563 - 5) = -825.156$

$y = (-825.156)^{\frac{1}{5}} = -3.831$

$$\delta y = \frac{[6(3.563)^5 - 25(3.563)^4] \times 0.001}{5(-3.831)^4}$$

$$= \frac{(3445.3263 - 4029.0559)0.001}{1077.0072}$$

$$= \frac{-0.5837296}{1077.0012} = -5.4199 \times 10^{-4}$$

$$= -5.4199 \times 10^{-4} \text{ to five significant figures.}$$

The new value of y is $-3.831 - 5.4199 \times 10^{-4}$ which is equal to -3.8315 to five significant figures.

L'Hôpital's Rule

Consider two functions $f(x)$ and $g(x)$ which represent two curves. If the curves intersect at P when $y = 0$ at $x = a$, then $f(a) = g(a) = 0$.

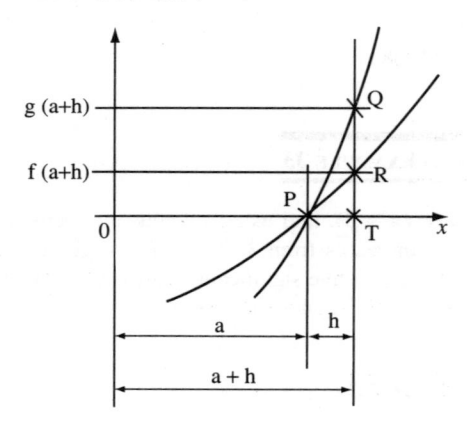

Fig. 4-I/24

If $x = a + h$, then at Q, $g(x) = g(a + h)$, at R, $f(x) = f(a + h)$

$QT = g(a + h)$, $RT = f(a + h)$

$\dfrac{QT}{RT} = \dfrac{g(a + h)}{f(a + h)}$ and dividing by PT, we have

$$\frac{\frac{g(a+h)}{h}}{\frac{f(a+h)}{h}} = \frac{\frac{QT}{PT}}{\frac{RT}{PT}} = \frac{\tan QPT}{\tan RPT}.$$

If $\lim\limits_{x \to a} \dfrac{g(x)}{f(x)} = \lim\limits_{h \to o} \dfrac{g(a+h)}{f(a+h)}$

$$= \lim_{h \to o} \frac{\tan QPT}{\tan RPT} = \lim \frac{g'(a)}{f'(a)}$$

provided $\dfrac{g(x)}{f(x)}$ is indeterminate, that is, $\dfrac{g(x)}{f(x)}$ is $\dfrac{0}{0}$ or $\dfrac{\infty}{\infty}$
and also provided $\dfrac{g'(a)}{f'(a)}$ is not indeterminate.

L'Hôpital's rule is only applied when $\dfrac{g(x)}{f(x)}$ is in the indeterminate form of $\dfrac{0}{0}$ or $\dfrac{\infty}{\infty}$.

WORKED EXAMPLE 34

Find the limit as $\Theta \to 0$ of $\dfrac{\sinh \Theta}{\Theta}$.

Solution 34

$$\lim_{\Theta \to 0} \frac{\sinh \Theta}{\Theta} = \frac{\sinh 0}{0} = \frac{0}{0}$$

this is indeterminate.

Applying L'Hôpital's rule, that is, differentiating numerator and denominator separately, we have

$$\lim_{\Theta \to 0} \frac{\sinh \Theta}{\Theta} = \lim_{\Theta \to 0} \frac{\cosh \Theta}{1} = \frac{\cosh 0}{1} = 1.$$

WORKED EXAMPLE 35

If $g(x) = e^x - \cos x$ and $f(x) = 5x$, find the limit as $x \to 0$ of the quotient $\dfrac{g(x)}{f(x)}$.

Solution 35

$$\lim_{x \to 0} \frac{e^x - \cos x}{5x} = \frac{e^0 - \cos 0}{5(0)} \text{ by direct}$$

substitution, then $\lim\limits_{x \to 0} \dfrac{e^x - \cos x}{5x} = \dfrac{1-1}{0} = \dfrac{0}{0}$.

Applying L'Hôpital's rule, that is, differentiating numerator and denominator separately, we have

$$\lim_{x \to 0} \frac{e^x - \cos x}{5x} = \lim_{x \to 0} \frac{e^x + \sin x}{5}$$

$$= \frac{e^0 + \sin(0)}{5} = \frac{1}{5}$$

therefore $\lim\limits_{x \to 0} \dfrac{e^x - \cos x}{5x} = \dfrac{1}{5}$.

Worked Example 36

Find the limit of the following as $x \to 0$:

(i) $\dfrac{\cos 5x - x \sin 3x - 1}{x^3}$

(ii) $\dfrac{4\cos x + x \sin 2x - 4}{x}$

(iii) $\dfrac{\cos 5x - \sin 7x - 1}{5 \sin 4x}$

(iv) $\dfrac{\cos x - \cos 3x}{2x^2}$

(v) $\dfrac{1 - x + \tan x - \cos x}{x^3}$

(vi) $\dfrac{\cos 3x - \cos x}{\sin 2x + \sin x}$

(vii) $\dfrac{\cos(x + \Theta) - \cos(2x - \Theta)}{2x}$

(viii) $\dfrac{\sin 3x}{3x}$

(ix) $\dfrac{\cos\left(x + \dfrac{\pi}{2}\right)}{\dfrac{\pi}{2} x}$

(x) $\dfrac{\tan(x - \alpha) + \tan \alpha}{x}$

Solution 36

(i) $\lim\limits_{x \to 0} \dfrac{\cos 5x - x \sin 3x - 1}{x^3} = \dfrac{0}{0}$ indeterminate

applying L'Hôpital's rule, that is, differentiate numerator and denominator separately

$\lim\limits_{x \to 0} \dfrac{-5 \sin 5x - \sin 3x - 3x \cos 3x}{3x^2} = \dfrac{0}{0}$

inderterminate, applying the rule again

$\lim\limits_{x \to 0} \dfrac{-25 \cos 5x - 3 \cos 3x - 3 \cos 3x + 9x \times \sin 3x}{6x}$

$= -\infty$

(ii) $\lim\limits_{x \to 0} \dfrac{4\cos x + x \sin 2x - 4}{x} = \dfrac{0}{0}$
indeterminate, applying the rule

$\lim\limits_{x \to 0} \dfrac{-4 \sin x + \sin 2x + 2x \cos 2x}{1} = \dfrac{0}{1} = 0$

(iii) $\lim\limits_{x \to 0} \dfrac{\cos 5x - \sin 7x - 1}{5 \sin 4x} = \dfrac{0}{0}$

indeterminate, applying the rule

$\lim\limits_{x \to 0} \dfrac{-5 \sin 5x - 7 \cos 7x}{20 \cos 4x} = -\dfrac{7}{20}$

(iv) $\lim\limits_{x \to 0} \dfrac{\cos x - \cos 3x}{2x^2} = \dfrac{0}{0}$

indeterminate, applying the rule

$\lim\limits_{x \to 0} \dfrac{-\sin x + 3 \sin 3x}{4x} = \dfrac{0}{0}$

indeterminate, applying the rule again

$\lim\limits_{x \to 0} \dfrac{-\cos x + 9 \cos 3x}{4} = 2$

(v) $\lim\limits_{x \to 0} \dfrac{1 - x + \tan x - \cos x}{x^3} = \dfrac{0}{0}$

indeterminate, applying the rule

$\lim\limits_{x \to 0} \dfrac{-1 + \sec^2 x + \sin x}{3x^2} = \dfrac{0}{0}$

indeterminate, applying the rule again

$\lim\limits_{x \to 0} \dfrac{2 \sec^2 x \tan x + \cos x}{6x} = \dfrac{1}{0} = \infty$

(vi) $\lim\limits_{x \to 0} \dfrac{\cos 3x - \cos x}{\sin 2x + \sin x} = \dfrac{0}{0}$

indeterminate, applying the rule

$\lim\limits_{x \to 0} \dfrac{-3 \sin 3x + \sin x}{2 \cos 2x + \cos x} = \dfrac{0}{3} = 0$

(vii) $\lim\limits_{x \to 0} \dfrac{\cos(x + \Theta) - \cos(2x - \Theta)}{2x} = \dfrac{0}{0}$

indeterminate, applying the rule

$\lim\limits_{x \to 0} \dfrac{-\sin(x + \Theta) + 2 \sin(2x + \Theta)}{2} = \dfrac{\sin \Theta}{2}$

(viii) $\lim\limits_{x \to 0} \dfrac{\sin 3x}{3x} = \dfrac{0}{0}$

indeterminate, applying the rule

$\lim\limits_{x \to 0} \dfrac{3 \cos 3x}{3} = 1$

(ix) $\lim_{x \to 0} \dfrac{\cos\left(x + \dfrac{\pi}{2}\right)}{\dfrac{\pi}{2}x} = \dfrac{0}{0}$

indeterminate, applying the rule

$\lim_{x \to 0} \dfrac{-\sin\left(x + \dfrac{\pi}{2}\right)}{\dfrac{\pi}{2}} = \dfrac{-1}{\dfrac{\pi}{2}} = -\dfrac{2}{\pi}$

(x) $\lim_{x \to 0} \dfrac{\tan(x - \alpha) + \tan \alpha}{x} = \dfrac{0}{0}$

indeterminate, applying the rule

$\lim_{x \to 0} \dfrac{\sec^2(x - \alpha) + \sec^2 \alpha}{1} = 2\sec^2 \alpha.$

Rate of Change

Any variable that changes with time is termed as the rate of change. A distance may change with time, an area may change with time, a volume may change with time, a velocity may change with time.

WORKED EXAMPLE 37

The area of a circle is given by $A = \pi r^2$.

Find the rate of change of A and that of the radius r.

Solution 37

$A = \pi r^2$, differentiating A with respect to r, we have
$\dfrac{dA}{dr} = 2\pi r$, dividing dA and dr by dt, we have

$\dfrac{\frac{dA}{dt}}{\frac{dr}{dt}} = 2\pi r = \dfrac{\text{the rate of change of } A}{\text{the rate of change of } r}$

$\dfrac{dA}{dt} = 2\pi r \dfrac{dr}{dt}.$

WORKED EXAMPLE 38

The surface area of a closed cylindrical can is given by $S = 2\pi rh + 2\pi r^2$. If h is constant, find the rates $\dfrac{dS}{dt}$ and $\dfrac{dr}{dt}$.

Solution 38

$S = 2\pi rh + 2\pi r^2$, differentiating with respect to r,
$\dfrac{dS}{dr} = 2\pi h + 4\pi r$, dividing both dS and dr by dt in order to introduce the respective rates $\dfrac{\frac{dS}{dt}}{\frac{dr}{dt}} = 2\pi h + 4\pi r.$

WORKED EXAMPLE 39

The surface and volume of a right circular cone are given by the expressions

$S = \pi r \left(h^2 + r^2\right)^{\frac{1}{2}} + \pi r^2$

$V = \dfrac{1}{3}\pi r^2 h.$

If $h = 100$ cm, constant, find the rates of change of S, V when $\dfrac{dr}{dt} = 1$ cm/s at $r = 10$ cm, to the nearest integer.

Solution 39

$\dfrac{dS}{dr} = \pi \left(h^2 + r^2\right)^{\frac{1}{2}} + \dfrac{1}{2}\pi r \left(h^2 + r^2\right)^{-\frac{1}{2}} 2r + 2\pi r$

$\dfrac{\frac{dS}{dt}}{\frac{dr}{dt}} = \pi \left(h^2 + r^2\right)^{\frac{1}{2}} + \dfrac{\pi r^2}{\left(h^2 + r^2\right)^{\frac{1}{2}}} + 2\pi r \quad \ldots(1)$

$\dfrac{dV}{dr} = \dfrac{2}{3}\pi rh, \quad \dfrac{\frac{dV}{dt}}{\frac{dr}{dt}} = \dfrac{2}{3}\pi rh \quad \ldots(2)$

From (1)

$\dfrac{dS}{dt} = \left[\pi \left(100^2 + 10^2\right)^{\frac{1}{2}} + \dfrac{\pi 10^2}{\left(100^2 + 10^2\right)^{\frac{1}{2}}} + 2\pi 10\right] 1$

$= 315.73 + 3.126 + 62.83 = 382 \text{ cm}^2/\text{s}$

From (2)

$\dfrac{dV}{dt} = \dfrac{2}{3}\pi rh \dfrac{dr}{dt} = \dfrac{2}{3}\pi(10)(100)(1) = 2094 \text{ cm}^3/\text{s}.$

WORKED EXAMPLE 40

The area of a circle is increasing at the rate of 5 cm^2 per second. Determine the rate of increase of the circumference of the circle when its radius is 6 cm.

Solution 40

$$A = \pi r^2 \qquad C = 2\pi r$$

$$\frac{dA}{dr} = 2\pi r \qquad \frac{dC}{dr} = 2\pi$$

$$\frac{\frac{dA}{dt}}{\frac{dr}{dt}} = 2\pi r \qquad \frac{\frac{dC}{dt}}{\frac{dr}{dt}} = 2\pi$$

$$2\pi r \frac{\frac{dC}{dt}}{\frac{dA}{dt}} = 2\pi \qquad \frac{dC}{dt} = \frac{dA}{dt} \cdot \frac{1}{r}$$

$$\frac{dA}{dt} = 5 \text{ cm}^2/\text{s}, \; \frac{dC}{dt} = 5 \times \frac{1}{6} = \frac{5}{6} \text{ cm/s}$$

WORKED EXAMPLE 41

A uniform soap bubble has a volume $V = \frac{4}{3}\pi r^3$; the rate of change of the volume to that of the radius is equal to 32π. Determine the radius of the bubble, and the rate of change of the surface area, if the volume increases by 100 cubic units per sec.

Solution 41

$$V = \frac{4}{3}\pi r^3 \qquad S = \frac{dV}{dr} = 4\pi r^2$$

$$\frac{\frac{dV}{dt}}{\frac{dr}{dt}} = 32\pi = 4\pi r^2, \quad r^2 = 8 \quad r = 2\sqrt{2} \text{ units}$$

$$S = 4\pi r^2 \qquad \frac{dS}{dr} = 8\pi r \qquad \frac{\frac{dS}{dt}}{\frac{dr}{dt}} = 8\pi r$$

$$\frac{dS}{dt} = \frac{8\pi r \frac{dV}{dt}}{32\pi} = \frac{2\sqrt{2}}{4}\frac{dV}{dt} = \frac{\sqrt{2}}{2}100 = 50\sqrt{2}.$$

Exercises 9

1. A horizontal trough is 5 m long and has a trapezoidal cross sectional area as shown in Fig. 4-I/25

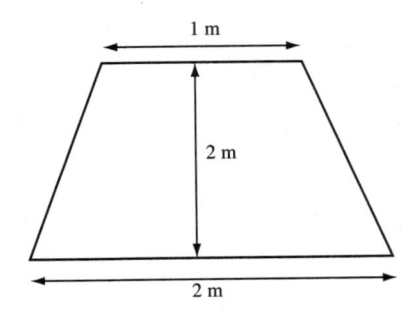

Fig. 4-I/25

Water runs into the trough at the rate of 1×10^{-3} m^3 per second. Find the rate at which the water level is rising when the height of the water is 1 m.

2. A spherical balloon has a radius of 10 m. Air is pumped into the balloon at the rate of 5×10^{-3} m^3 per second. Determine the rate at which the radius and the surface area increase.

3. The surface area S and volume V of a solid sphere are changing with respect to time t when it is uniformly heated. When its surface area is increasing at a rate of 1 mm^2/s, determine the rate at which the volume is increasing when its radius is 10 cm.

4. The volume of an expanding sphere is increasing at the rate of 100 mm^3/s. Determine the rate at which the radius of the sphere is increasing at the instant when the radius is 75 mm.

5. The volume of a certain bowl is changing with its height according to the equation $\frac{dV}{dh} = 3e^{3h} + 5e^h + 1$.

   If the rate at which the height changes is 3 cm/s, determine the rate at which the volume changes when $h = 1$ m.

6. An ellipse has a cartesian equation $4x^2 + 9y^2 = 36$; if x is increasing at the rate of 0.1 cm/sec, find the rate of decrease of y when $x = 2$.

7. The volume of a cap of a sphere of height h is given by $V = \frac{\pi h^2}{3}(3r - h)$, where r is the radius. If $\frac{dr}{dt} = 5$ cm/sec and $h = 20$ cm $=$ constant, find $\frac{dV}{dt}$ when $r = 50$ cm.

8. An equilateral triangle has a side of 10 cm and its perimeter increases at the rate of 1 cm/s; find the rate of increase of the area. If each side increases to 10.1 cm, find the corresponding increase in the area.

9. Find the limits

   (i) $\lim_{x \to 0} \frac{\sin 5x}{5x}$

   (ii) $\lim_{x \to 0} \frac{\tan kx}{x}$

   (iii) $\lim_{x \to 0} \frac{2 \sin^{-1} x}{3x}$

   (iv) $\lim_{x \to 0} \frac{1 - \cos x}{x^2}$

(v) $\lim\limits_{x\to 0}\left(\dfrac{1}{\sin x}-\dfrac{1}{\tan x}\right)$

(vi) $\lim\limits_{x\to \frac{\pi}{2}}\dfrac{\cos x}{\sqrt{(1-\sin x)^{\frac{2}{3}}}}$

(vii) $\lim\limits_{x\to 0}\dfrac{1-\cos^3 x}{x\sin 2x}$

(viii) $\lim\limits_{x\to \frac{\pi}{2}}\dfrac{1-\sin x}{\left(\dfrac{\pi}{2}-x\right)^2}$

(ix) $\lim\limits_{x\to \pi}\dfrac{\sin 3x}{\sin 2x}$

(x) $\lim\limits_{x\to 1}\dfrac{e^x-e}{x-1}$

(xi) $\lim\limits_{x\to 0}\dfrac{e^{x^2}-\cos x}{x^2}$

(xii) $\lim\limits_{x\to 0}\dfrac{e^x-e^{-x}}{\sin x}$

(xiii) $\lim\limits_{x\to 0}\dfrac{\ln\cos x}{x^2}$

(xiv) $\lim\limits_{x\to 0}\dfrac{1-\cos(1-\cos x)}{x^3}$

(xv) $\lim\limits_{x\to 0}\dfrac{1-\sin(1+\sin x)}{x^4}.$

10

Newton-Raphson Approximation

If x_n is an approximation to a root of $f(x) = 0$ then usually a better approximation is given by x_{n+1}

$$x_{n+1} = x_n - \frac{f(x_n)}{f'(x_n)}$$

Consider a curve that intersects the x-axis at $x = x_n$ as shown in Fig. 4-I/26; the corresponding value of y is $f(x_n)$ which is approximately zero. A better approximation can be achieved that will make the value of y nearer to zero, let this point be at Q which is very close to P; the gradient of the chord PQ is given by

$$\frac{\delta y}{\delta x} = \frac{f(x_{n+1}) - f(x_n)}{x_{n+1} - x_n}$$

As δx approaches zero, $\dfrac{\delta y}{\delta x}$ approaches the gradient of the tangent at P, according to the idea of the limit explained previously.

As $\delta x \to 0$, $\dfrac{\delta y}{\delta x} \to \dfrac{dy}{dx} = f'(x_n)$ and $f(x_{n+1}) \to 0$

$$\frac{dy}{dx} = f'(x_n) = -\frac{f(x_n)}{x_{n+1} - x_n}$$

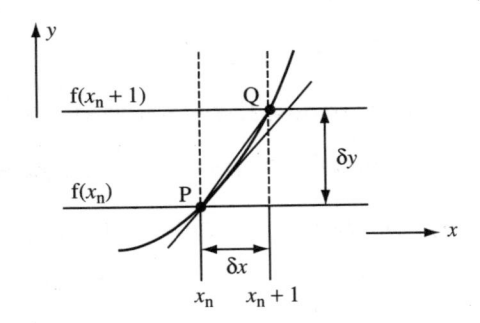

Fig. 4-I/26

$$x_{n+1} - x_n = -\frac{f(x_n)}{f'(x_n)}$$

$$x_{n+1} = x_n - \frac{f(x_n)}{f'(x_n)}$$

Worked Example 42

Plot the graph of the cubic function $y = 2x^3 - 3x^2 - 11x - 12$ between the values $x = -4$ and $x = 4$, at intervals of 0.5. Give your answers correct to two decimal places, for the roots of the equation $2x^3 - 3x^2 - 11x + 8 = 0$.

Solution 42

x	-4	-3.5	-3.0	-2.5
	0.5	1	1.5	2
	-12	-12	-12	-12
	-12	-12	-12	-12
$-11x$	44	38.5	33	27.5
	-5.50	-11	-16.5	-22
$-3x^2$	-48	-36.75	-27	-18.75
	-0.75	-3	-6.75	-12
$2x^3$	-128	-85.75	-54	-31.25
	0.25	2	6.75	16
y	-144	-96	-60	-34.5
	-18.00	-24	-28.5	-30

contd ...

x	-2.0	-1.5	-1.0	-0.5	0	
		2.5	3	3.5	4	
-12	-12	-12	-12	-12	-12	-12
		-12	-12	-12	-12	
$-11x$	22	16.5	11	5.5	0	
		-27.5	-33	-38.5	-44	
$-3x^2$	-12	-6.75	-3	-0.75	0	
		-18.75	-27	-36.25	-48	
$2x^3$	-16	-6.75	-2	-0.25	0	
		31.25	54	85.75	128	
y	-18	-9	-6	-7.5	-12	
		-27	-18	-1.00	24	

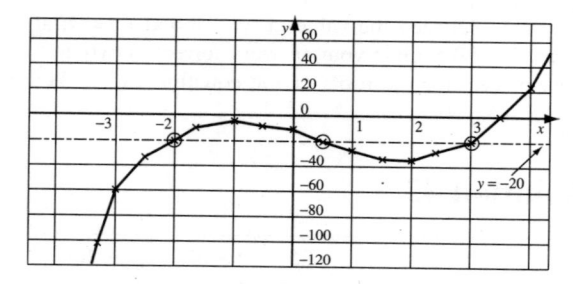

Fig. 4-I/27

Use the following scales on the x-axis, 1 cm = 0.5 and the y-axis, 1 cm = 10. y is plotted against x and a smooth curve is obtained with a maximum point at $x = -1$ and a minimum point at $x = 2$; the graph cuts the x-axis at $x = 3.55$ and the y-axis at $y = -12$. In order to solve the equation $2x^3 - 3x^2 - 11x + 8 = 0$, we make $y = -20$, a horizontal line intersects the graph at $x = -2.08$, $x = 0.68$ and $x = 2.93$.

$$y = 2x^3 - 3x^2 - 11x - 12 = -20.$$

Let $f(x) = 2x^3 - 3x^2 - 11x + 8$ ...(1)

We would like to find better approximations for the roots $x_{n_1} = -2.08$, $x_{n_2} = 0.68$ and $x_{n_3} = 2.93$.

Let us apply the Newton-Raphson method of approximation.

$$x_{n_1+1} = x_{n_1} - \frac{f(x_{n_1})}{f'(x_{n_1})}$$

$$f(x_{n_1}) = 2(-2.08)^3 - 3(-2.08)^2 - 11(-2.08) + 8$$
$$= -17.998 - 12.98 + 22.88 + 8 = -0.098.$$

Differentiating equation (1), we have

$$f'(x) = 6x^2 - 6x - 11, \text{ for } x = -2.08$$

$$f'(x_{n_1}) = 6(-2.08)^2 - 6(-2.08) - 11$$
$$= 25.9584 + 12.48 - 11 = 27.4384.$$

A better approximation is x_{n_1+1}

$$x_{n_1+1} = x_{n_1} - \frac{f(x_{n_1})}{f'(x_{n_1})} = -2.08 - \frac{(-0.098)}{27.4384}$$

$$= -2.08 + 0.0035716 = -2.076.$$

This may be further improved if the method is repeated.

$$f(x_{n_1+1}) = 2(-2.076)^3 - 3(-2.076)^2$$
$$- 11(-2.076) + 8$$
$$= -17.89419 - 12.929328 + 22.836 + 8$$
$$= 0.012482 = 0.0125 \text{ to four decimal places}$$

$$f'(-2.076) = 6(-2.076)^2 - 6(-2.076) - 11$$
$$= 25.858656 + 12.456 - 11 = 27.314656$$

$$(x_{n_1+2}) = (x_{n_1+1}) - \frac{f(x_{n_1+1})}{f'(x_{n_1+1})} = -2.076 - \frac{0.0125}{27.314656}$$

$$= -2.076 - 0.00045762978$$

$x_{n_1+2} = -2.0765$

$x_{n_1+2} = -2.077$ to four significant figures

$= -2.08$ to three significant figures

$= -2.1$ to two significant figures.

Let us apply again the Newton-Raphson method of approximation for the second root, $x_{n_2} = 0.68$

$$f(x) = 2x^3 - 3x^2 - 11x + 8$$

$$f'(x) = 6x^2 - 6x - 11$$

$f(x_{n_2}) = 2(0.68)^3 - 3(0.68)^2 - 11(0.68) + 8$

$= 0.628864 - 1.3872 - 7.48 + 8 = -0.238$

$f'(x_{n_2}) = 6(0.68)^2 - 6(0.68) - 11 = -12.3056$

$x_{n_2+1} = x_{n_2} - \dfrac{f(x_{n_2})}{f'(x_{n_2})} = 0.68 - \dfrac{(-0.238)}{-12.3056}$

$x_{n_2+1} = 0.661 = 0.66$ to two decimal places.

Let us apply finally the Newton-Raphson method to the third root, $x_{n_3} = 2.93$.

$f(x) = 2x^3 - 3x^2 - 11x + 8$

$f'(x) = 6x^2 - 6x - 11$,

$f'(x_{n_3}) = 6(2.93)^2 - 6(2.93) - 11 = 22.9294$

$f(x_{n_3}) = 2(2.93)^3 - 3(2.93)^2 - 11(2.93) + 8$

$= 50.307514 - 25.7547 - 32.23 + 8$

$= 0.322814.$

A better approximation is as follows:-

$x_{n_3+1} = x_{n_3} - \dfrac{f(x_{n_3})}{f'(x_{n_3})}$

$= 2.93 - \dfrac{0.322814}{22.9294} = 2.9159214$

$= 2.92$ to three significant figures.

The roots of the cubic equation $2x^3 - 3x^2 - 11x + 8 = 0$ are now improved to the values -2.08, 0.66 and 2.92 to three significant figures.

WORKED EXAMPLE 43

Taking $x = 43.5°$ as a first approximation to a root of the equation

$\sin x + \sin \dfrac{x}{2} - 1 = 0$, apply the Newton-Raphson procedure to find a further approximation giving your answer to 3 significant figures.

Solution 43

$f(x) = \sin x + \sin \dfrac{x}{2} - 1$

$f'(x) = \cos x + \dfrac{1}{2} \cos \dfrac{1}{2} x$

$x_n = 43.5°$, it is required to find x_{n+1}, a better approximation.

$f(x_n) = \sin 43.5° + \sin \dfrac{1}{2}(43.5°) - 1$

$= 0.6883545 + 0.3705574 - 1 = 0.0589119$

$f'(x_n) = \cos 43.5° + \dfrac{1}{2} \cos \dfrac{1}{2}(43.5°) = 1.1897791$

$x_{n+1} = x_n - \dfrac{f(x_n)}{f'(x_n)} = 43.5 \times \dfrac{\pi}{180} - \dfrac{0.0589119}{1.1897791}$

$= 0.759218 - 0.0495149$

$= 0.709703^c \times \dfrac{180}{\pi} = 40.662987° = 40.7°$

WORKED EXAMPLE 44

Taking $x = 0.324^c$ as a first approximation to a root of the equation $2\cos\left(2x - \dfrac{2\pi}{3}\right) - x = 0$, apply the Newton-Raphson procedure to find two further approximations giving your answers to 3 decimal places.

Solution 44

$f(x) = 2\cos\left(2x - \dfrac{2\pi}{3}\right) - x$

$f'(x) = -4\sin\left(2x - \dfrac{2\pi}{3}\right) - 1.$

When $x_n = 0.324^c$ (note the angle is in radians)

$f(x_n) = 2\cos\left(2 \times 0.324 - \dfrac{2\pi}{3}\right)^c - 0.324^c$

$= -0.0758387$

$f'(x_n) = -4\sin\left(2 \times 0.324 - \dfrac{2\pi}{3}\right)^c - 1 = 2.9690886.$

A better approximation is found as follows:-

$x_{n+1} = x_n - \dfrac{f(x_n)}{f'(x_n)} = 0.324 - \dfrac{(-0.0758387)}{2.9690886}$

$= 0.3495427 = 0.350^c$ to three decimal places.

A better still approximation is found as follows:-

$$x_{n+2} = x_{n+1} - \frac{f(x_{n+1})}{f'(x_{n+1})}$$

$$f(x_{n+1}) = 2\cos\left(2 \times 0.350 - \frac{2\pi}{3}\right) - 0.350$$

$$= 0.00097557814$$

$$f'(x_{n+1}) = -4\sin\left(2 \times 0.350 - \frac{2\pi}{3}\right) - 1$$

$$= 2.9379264$$

$$x_{n+2} = x_{n+1} - \frac{f(x_{n+1})}{f'(x_{n+1})} = 0.350 - \frac{0.00097557814}{2.9379264}$$

$$= 0.3496679 = 0.350^c \text{ to three decimal places.}$$

The two answers are correct to three decimal places.

WORKED EXAMPLE 45

The graphs of $y = \tan^{-1} x$ and $\frac{x}{5} + \frac{y}{6} = 1$ are drawn giving points of intersections, to the first approximations, of $(3.8, 1.313^c)$ and $(1.55, 4.14^c)$.

Find better approximations to these solutions by applying the Newton-Raphson formula.

Solution 45

$$y = \tan^{-1} x, \quad \frac{x}{5} + \frac{y}{6} = 1$$

$$y = \tan^{-1} x = 6\left(1 - \frac{x}{5}\right) = 6 - \frac{6}{5}x$$

$$5\tan^{-1} x + 6x - 30 = 0$$

$$f(x) = 5\tan^{-1} x + 6x - 30$$

$$f'(x) = \frac{5}{1 + x^2} + 6$$

$$f(x_{n_1}) = 5\tan^{-1} x_{n_1} + 6x_{n_1} - 30 \text{ and}$$

$$f'(x_{n_1}) = \frac{5}{1 + x_{n_1}^2} + 6.$$

For $(3.8, 1.313^c)$, that is, $x_{n_1} = 3.8$ and $\tan^{-1} x_{n_1} = 1.313$,

$$f(x_{n_1}) = 5 \times 1.313 + 6 \times 3.8 - 30$$

$$= 29.365 - 30 = -0.635$$

$$f'(x_{n_1}) = \frac{5}{1 + x_{n_1}^2} + 6 = \frac{5}{1 + 3.8^2} + 6 = 6.3238342$$

$$x_{n_1+1} = x_{n_1} - \frac{f(x_{n_1})}{f'(x_{n_1})} = 3.8 - \frac{(-0.635)}{6.3238342} = 3.90.$$

For $(1.55, 4.14^c)$ that is, $x_{n_2} = 1.55$ and $\tan^{-1} x_{n_2} = 4.14^c$

$$f(x_{n_2}) = 5\tan^{-1} x_{n_2} + 6x_{n_2} - 30$$
$$= 5 \times 4.14 + 6 \times 1.55 - 30 = 0$$

$$f'(x_{n_2}) = \frac{5}{1 + x_{n_2}^2} + 6 = \frac{5}{1 + 1.55^2} + 6 = 7.4695077$$

$$x_{n_2+1} = x_{n_2} - \frac{f(x_{n_2})}{f'(x_{n_2})} = 1.55 - \frac{0}{7.4695077} = 1.55$$

Therefore $x_{n_2+1} = 1.55$.

The two better approximations are:-

$x_{n_1+1} = 3.90$ and $x_{n_2+1} = 1.55$.

Exercises 10

1. Show graphically, or otherwise, that the cubic equations

   (i) $x^3 + 3x + 3 = 0$

   (ii) $x^3 + 3x + 28 = 0$

   have only one real root, and prove that this root lies between -0.7 and -0.9 for (i) and between -2.6 and -2.8 for (ii).

   By taking $x = -0.75$ for (i) and $x = -2.7$ for (ii) as first approximations to these roots, use the Newton-Raphson method to obtain improved solutions giving your answer to 3 significant figures.

2. Find graphically the solution of the equation $f(x) = 2x - e^{-x} = 0$. (The graphs of $y = 2x$ and $y = e^{-x}$ are plotted, the intersection gives the solution $x = 0.35$).

   Taking $x = 0.35$ as a first approximation apply the Newton-Raphson procedure to the root of the equation $f(x) = 0$ giving your answer to 4 decimal places.

3. Solve the equations $3x - e^x = 0$, $4x - e^x = 0$ graphically and use the Newton-Raphson method to obtain improved solutions.

4. Taking $x = 1.414^c$ as a first approximation to a root of the equation $\left(\cos 2x - \frac{2\pi}{3}\right) = \frac{x}{2}$ apply the Newton-Raphson procedure to find two further approximations giving your answers to three decimal places.

11
Maclaurin's Expansions

Power Series

The following are some examples of infinite series, series in which the number of terms is allowed to increase without a limit.

1. $(1+x)^n = 1 + nx + \dfrac{n(n-1)}{2!}x^2$
   $+ \dfrac{n(n-1)(n-2)}{3!}x^3 + \ldots$
   $+ \dfrac{n(n-1)\ldots(n-r+1)}{r!}x^r + \ldots$
   (n is not a positive integer, $|x| < 1$).

2. $e^x = \exp x = 1 + x + \dfrac{x^2}{2!} + \dfrac{x^3}{3!} + \ldots + \dfrac{x^r}{r!} + \ldots$
   for all values of x

3. $\ln(1+x) = x - \dfrac{x^2}{2} + \dfrac{x^3}{3} - \ldots + (-1)^{r+1}\dfrac{x^r}{r} + \ldots$ $(-1 < x \le 1)$

4. $\cos x = 1 - \dfrac{x^2}{2!} + \dfrac{x^4}{4!} - \ldots + (-1)^r \dfrac{x^{2r}}{(2r)!} + \ldots$ for all values of x

5. $\sin x = x - \dfrac{x^3}{3!} + \dfrac{x^5}{5!} - \ldots + (-1)^r \dfrac{x^{2r+1}}{(2r+1)!} + \ldots$
   for all values of x

6. $\cosh x = 1 + \dfrac{x^2}{2!} + \dfrac{x^4}{4!} + \ldots + \dfrac{x^{2r}}{(2r)!} + \ldots$ for all values of x

7. $\sinh x = x + \dfrac{x^3}{3!} + \dfrac{x^5}{5!} + \ldots + \dfrac{x^{2r+1}}{(2r+1)!} + \ldots$ for all values of x.

WORKED EXAMPLE 46

Use the expansions of e^x and e^{-x} to obtain the following expansions:-

(i) $\cosh x$

(ii) $\sinh x$

(iii) $\cos x$

(iv) $\sin x$.

Solution 46

(i) $\cosh x = \dfrac{1}{2}\left[e^x + e^{-x}\right]$

$= \dfrac{1}{2}\left[1 + x + \dfrac{x^2}{2!} + \ldots + 1 - x + \dfrac{x^2}{2!} - \ldots\right]$

$= \dfrac{1}{2}\left[2 + 2\dfrac{x^2}{2!} + 2\dfrac{x^4}{4!} + \ldots\right] = 1 + \dfrac{x^2}{2!} + \dfrac{x^4}{4!} + \ldots$

(ii) $\sinh x = \dfrac{1}{2}\left[e^x - e^{-x}\right]$

$= \dfrac{1}{2}\left[1 + x + \dfrac{x^2}{2!} + \ldots - \left(1 - x + \dfrac{x^2}{2!} - \ldots\right)\right]$

$= \dfrac{1}{2}\left[2x + 2\dfrac{x^3}{3!} + \ldots\right] = x + \dfrac{x^3}{3!} + \ldots$

(iii) $\cos x = \dfrac{1}{2}\left[e^{ix} + e^{-ix}\right]$

$= \dfrac{1}{2}\left[\left(1 + ix + \dfrac{i^2 x^2}{2!} + \ldots\right)\right.$

$$+ \left(1 - ix + \frac{i^2x^2}{2!} - \cdots\right)\bigg]$$

$$= 1 - \frac{x^2}{2!} + \frac{x^4}{4!} - \cdots$$

(iv) $\sin x = \dfrac{e^{ix} - e^{-ix}}{2}$

$$= \frac{1}{2}\left[\left(1 + ix + \frac{i^2x^2}{2!} + \cdots\right)\right.$$

$$\left. - \left(1 - ix + \frac{i^2x^2}{2!} - \cdots\right)\right]$$

$$= x - \frac{x^3}{3!} + \frac{x^5}{5!} - \cdots$$

WORKED EXAMPLE 47

Use the expansions of $\sin x$ and $\cos x$ to obtain the following expansions:-

(i) $\tan x$ and

(ii) $\tanh x$, as far as the term in x^5.

Solution 47

(i) $\tan x = \dfrac{\sin x}{\cos x} = \dfrac{x - \frac{x^3}{3} + \frac{x^5}{5} - \cdots}{1 - \frac{x^2}{2} + \frac{x^4}{4} - \cdots}$

Using long division, we have

$$1 - \frac{x^2}{2!} + \frac{x^4}{4!} - \cdots \overline{\left| x - \frac{x^3}{3!} + \frac{x^5}{5!} - \cdots \right.} \quad \overline{x + \frac{1}{3}x^3 + \frac{2}{15}x^5}$$

$$\underline{x - \frac{x^3}{2!} + \frac{x^5}{4!} - \cdots}$$

$$\left(-\frac{1}{6} + \frac{1}{2}\right)x^3 + \left(\frac{1}{120} - \frac{1}{24}\right)x^5$$

$$= \frac{1}{3}x^3 - \frac{1}{30}x^5$$

$$\underline{\frac{1}{3}x^3 - \frac{1}{6}x^5}$$

$$\left(-\frac{1}{30} + \frac{1}{6}\right)x^5 = \frac{2}{15}x^5$$

$$\boxed{\tan x = x + \frac{1}{3}x^3 + \frac{2}{15}x^5}$$

(ii) $\tanh x = \dfrac{\sinh x}{\cosh x} = \dfrac{x + \frac{x^3}{3}! + \frac{x^5}{5}! + \cdots}{1 + \frac{x^2}{2}! + \frac{x^4}{4}! + \cdots}$

using long division, we have

$$1 + \frac{x^2}{2!} + \frac{x^4}{4!} + \cdots \overline{\left| x + \frac{x^3}{3!} + \frac{x^5}{5!} + \cdots \right.} \quad \overline{x - \frac{1}{3}x^3 + \frac{2}{15}x^5}$$

$$\underline{x + \frac{x^3}{2!} + \frac{x^5}{4!} + \cdots}$$

$$\left(\frac{1}{6} - \frac{1}{2}\right)x^3 + \left(\frac{1}{120} - \frac{1}{24}\right)x^5$$

$$= -\frac{1}{3}x^3 - \frac{1}{30}x^5$$

$$\underline{-\frac{1}{3}x^3 - \frac{1}{6}x^5}$$

$$\left(-\frac{1}{30} + \frac{1}{6}\right)x^5 = \frac{2}{15}x^5$$

$$\boxed{\tanh x = x - \frac{1}{3}x^3 + \frac{2}{15}x^5.}$$

WORKED EXAMPLE 48

Expand $e^{\sin x}$ as far as the term in x^4.

Solution 48

$$e^{\sin x} = e^{x - \frac{x^3}{3!} + \frac{x^5}{5!} - \cdots} = e^x e^{-\frac{x^3}{3!}} e^{\frac{x^5}{5!}} e^{-\frac{x^7}{7!}} \cdots$$

$$= \left(1 + x + \frac{x^2}{2!} + \frac{x^3}{3!} + \frac{x^4}{4!}\right)$$

$$\left(1 - \frac{x^3}{3!}\right)\left(1 + \frac{x^5}{5!}\right)\left(1 - \frac{x^7}{7!}\right)$$

$$= 1 + x + \frac{x^2}{2!} + \frac{x^3}{3!} + \frac{x^4}{4!} - \frac{x^3}{3!} - \frac{x^4}{3!}$$

$$= 1 + x + \frac{1}{2}x^2 + \left(\frac{1}{24} - \frac{1}{6}\right)x^4$$

$$= 1 + x + \frac{1}{2}x^2 - \frac{1}{8}x^4$$

Taylor's Theorem States

$$f(x+h) = f(x) + hf'(x) + \frac{h^2}{2!}f''(x) + \frac{h^3}{3!}f'''(x) + \ldots$$

It is required to find a formula for the expansion $f(x+h)$ in a series of ascending powers of h.

Let $f(x+h) = a_0 + a_1 h + a_2 h^2 + a_3 h^3 + \ldots + a_n h^n + \ldots$

Differentiating with respect to h and regarding x as a constant

$$f'(x+h) = a_1 + 2a_2 h + 3a_3 h^2 + \ldots + n a_n h^{n-1} + \ldots$$

$$f''(x+h) = 2a_2 + 3 \times 2a_3 h + \ldots + n(n-1)a_n h^{n-2} + \ldots$$

$$f'''(x+h) = 3 \times 2 \times 1 a_3 + \ldots$$
$$+ n(n-1)(n-2)a_n h^{n-3} + \ldots$$

$$f^{(n)}(x+h) = n(n-1)(n-2) \ldots 3 \times 2 a_n + \text{power of } h$$

$f'(x) = a_1, f''(x) = 2a_2, f'''(x) = 3 \times 2 a_3$, and, in general $f^{(n)}(x) = n! a_n$.

Thus $a_0 = \dfrac{f(x)}{0!}, a_1 = \dfrac{f'(x)}{1!}, a_2 = \dfrac{f''(x)}{2!},$
$a_3 = \dfrac{f'''(x)}{3!}, \ldots, a_n = \dfrac{f^n(x)}{n!}.$

Therefore, Taylor's Theorem

$$\boxed{f(x+h) = f(x)\frac{h^0}{0!} + f'(x)\frac{h^1}{1!} + f''(x)\frac{h^2}{2!} + f'''(x)\frac{h^3}{3!} + \ldots f^{(n)}(x)\frac{h^n}{n!} + \ldots}$$

Putting $x = 0$,

$$f(h) = f(0)\frac{h^0}{0!} + f'(0)\frac{h^1}{1!} + f''(0)\frac{h^2}{2!} + f'''(0)\frac{h^3}{3!} + \ldots + f^{(n)}(0)\frac{h^n}{n!} + \ldots$$

which is Maclaurin's Theorem, with h written for x. With x and h interchanged Taylor's Theorem can be written

$$\boxed{f(x+h) = f(h)\frac{x^0}{0!} + f'(h)\frac{x^1}{1!} + f''(h)\frac{x^2}{2!} + f'''(h)\frac{x^3}{3!} + \ldots + f^{(n)}(h)\frac{x^n}{n!} + \ldots}$$

WORKED EXAMPLE 49

Expand $\sin(x+h)$ in powers of h by Taylor's Theorem, and deduce the value of $\sin 31°$ correct to 5 decimal places.

Solution 49

Let $f(x+h) = \sin(x+h)$
$$f(x) = \sin x$$
$$f'(x) = \cos x$$
$$f''(x) = -\sin x$$

Taylor's Theorem gives

$$\sin(x+h) = \sin x + h \cos x + h^2 \frac{(-\sin x)}{2!}$$
$$+ h^3 \frac{(-\cos x)}{3!} + \ldots$$

or $\sin(x+h) = \sin x + h \cos x - \dfrac{h^2}{2!}\sin x - \dfrac{h^3}{3!}\cos x + \ldots$

Let $x = 30° = \dfrac{\pi}{6}$

$$\sin(30° + h) = \sin 30° + h \cos 30°$$
$$- \frac{h^2}{2} \sin 30° - \frac{h^3}{3!} \cos 30° + \ldots$$
$$= \frac{1}{2} + h \frac{\sqrt{3}}{2} - \frac{h^2}{2} \cdot \frac{1}{2} - \frac{1}{6} h^3 \frac{\sqrt{3}}{2} + \ldots$$

If now $h = 1° = 0.01745^c$, $h^2 = 0.000304$, and $h^3 = 0.000005$

$\sin 31° \approx 0.5 + 0.866 \times 0.01745 - 0.25 \times 0.000304$

$\sin 31° \approx 0.51504.$

From the calculator $\sin 31° = 0.515038 \approx 0.51504$ correct to 5 decimal places.

Approximate Solution of Equations

When an approximate solution of an equation has been found by a graphical method, successive nearer approximations can be found by Taylor's Theorem.

Let $f(x) = 0$ be the equation, and let $x = a$ be the approximate solution found, so that $f(a)$, though not exactly zero, is quite small.

Let $x = a + h$ be the actual value of the root, where h is necessarily small. Then

58 — GCE A level

$f(a+h) = 0$, and by Taylor's Theorem, $f(a) + hf'(a) + \dfrac{h^2}{2}f''(a) + \ldots = 0$. Thus neglecting h^2 and higher powers, $f(a) + hf'(a) \approx 0$,

so that $h \approx -\dfrac{f(a)}{f'(a)}$, and the nearer approximation is

$$\boxed{x \approx a - \dfrac{f(a)}{f'(a)}}$$

This process may be repeated as necessary to give a still nearer approximation. This is the Newton-Raphson's approximation.

$$x_{n+1} = x_n - \dfrac{f(x_n)}{f'(x_n)}.$$

WORKED EXAMPLE 50

Expand $\cos(x+h)$ in powers of h by Taylor's Theorem. By putting $x = \dfrac{\pi}{3}$ and $h = \left(1° \times \dfrac{\pi}{180}\right)^c$, deduce the value of $\cos 61°$ correct to 5 decimal places.

Solution 50

Let $f(x+h) = \cos(x+h)$

$f(x) = \cos x$

$f'(x) = -\sin x$

$f''(x) = -\cos x$

$f'''(x) = \sin x$

Taylor's Theorem gives

$\cos(x+h) = \cos x + h(-\sin x)$

$\qquad + \dfrac{h^2}{2!}(-\cos x) + \dfrac{h^3}{3!}\sin x$

or $\cos(x+h) = \cos x - h\sin x - \dfrac{h^2}{2}\cos x + \dfrac{h^3}{6}\sin x.$

Let $x = 60° = \dfrac{\pi}{3}$

$\cos(60° + h) = \cos 60° - h\sin 60°$

$\qquad - \dfrac{h^2}{2}\cos 60° + \dfrac{h^3}{6}\sin 60°$

$\qquad = 0.5 - h\,0.866 - \dfrac{h^2}{2}0.5 + \dfrac{h^3}{6}0.866$

If now $h = 1° = 0.0174532^c$, $h^2 = 0.00030461742$, $h^3 = 0.0000053165769$

$\cos 61° = 0.5 - 0.0174532 \times 0.866 - 0.00030461742$

$\qquad \times 0.25 + 7.6735927 \times 10^{-7}$

$\qquad = 0.4848101 \approx 0.48481$

correct to 5 decimal places.

From the calculator $\cos 61° \approx 0.48481$ correct to 5 decimal places.

WORKED EXAMPLE 51

Obtain the expansion $\tan\left(\dfrac{\pi}{4} + x\right) = 1 + 2x + 2x^2 + \dfrac{8}{3}x^3 + \dfrac{10}{3}x^4 + \ldots$ and use it to calculate the value of $\tan 46°$ to 5 decimal places.

Solution 51

Let $f(x+h) = \tan\left(\dfrac{\pi}{4} + x\right)$

$f(x) = \tan x$

$f'(x) = \sec^2 x$

$f''(x) = 2\sec^2 x \tan x$

$f'''(x) = 4\sec^2 x \tan^2 x + 2\sec^4 x$

Taylor's Theorem gives

$\tan\left(\dfrac{\pi}{4} + x\right) = \tan x + h\sec^2 x + \dfrac{h^2}{2!}2\sec^2 x \tan x$

$\qquad + \dfrac{h^3}{3!}\left(4\sec^2 x \tan^2 x + 2\sec^4 x\right)$

Let $x = \dfrac{\pi}{4} = 45°$,

$\tan\left(\dfrac{\pi}{4} + x\right) = \tan\dfrac{\pi}{4} + \dfrac{\pi}{180}\sec^2\dfrac{\pi}{4}$

$\qquad + \left(\dfrac{\pi}{180}\right)^2 \sec^2\dfrac{\pi}{4}\tan\dfrac{\pi}{4} + \left(\dfrac{\pi}{180}\right)^3 \dfrac{1}{6}$

$\qquad \left(4\sec^2\dfrac{\pi}{4}\tan^2\dfrac{\pi}{4} + 2\sec^4\dfrac{\pi}{4}\right)$

$\tan 46° = 1 + 0.0349065 + 6.0923484 \times 10^{-4} + 1.4177538 \times 10^{-5} = 1.0355299 \approx 1.03553$ correct to 5 decimal places.

From the calculator $\tan 46° = 1.03553$ correct to 5 decimal places.

Worked Example 52

Prove, by Taylor's Theorem, that

$$\sin^{-1}(x+h) = \sin^{-1}x + \frac{h}{\sqrt{1-x^2}} + \frac{x}{(1-x^2)^{\frac{3}{2}}}\frac{h^2}{2!}$$
$$+ \frac{1+2x^2}{(1+x^2)^{\frac{5}{2}}}\frac{h^3}{3!} + \ldots$$

and hence evaluate $\sin^{-1}(0.501)$.

Solution 52

Let $f(x+h) = \sin^{-1}(x+h)$

$$f(x) = \sin^{-1}x$$

$$f'(x) = \frac{1}{(1-x^2)^{\frac{1}{2}}} = (1-x^2)^{-\frac{1}{2}}$$

$$f''(x) = -\frac{1}{2}(-2x)(1-x^2)^{-\frac{3}{2}}$$
$$= x(1-x^2)^{-\frac{3}{2}}$$

$$f'''(x) = (1-x^2)^{-\frac{3}{2}}$$
$$+ x\left(-\frac{3}{2}\right)(-2x)(1-x^2)^{-\frac{5}{2}}$$
$$= (1-x^2)^{-\frac{3}{2}} + \frac{3x^2}{(1-x^2)^{\frac{5}{2}}}$$
$$= \frac{1-x^2+3x^2}{(1-x^2)^{\frac{5}{2}}} = \frac{1+2x^2}{(1-x^2)^{\frac{5}{2}}}$$

Taylor's Theorem gives

$$\sin^{-1}(x+h) = \sin^{-1}x + \frac{h}{\sqrt{1-x^2}} + \frac{x}{(1-x^2)^{\frac{3}{2}}}\frac{h^2}{2!}$$
$$+ \frac{1+2x^2}{(1-x^2)^{\frac{5}{2}}}\frac{h^3}{3!} + \ldots$$

Putting $x = 0.5$ and $h = 0.001$, we have

$$\sin^{-1}(0.501) = \sin^{-1}0.5 + \frac{0.001}{\sqrt{1-0.25}}$$
$$+ \frac{0.5}{(1-0.25)^{\frac{3}{2}}}\frac{0.001^2}{2}$$
$$+ \frac{1+2(0.25)}{(1-0.25)^{\frac{5}{2}}} \times \frac{0.001^3}{6}$$
$$= \frac{\pi}{6} + 1.1547 \times 10^{-3} + 7.6980036$$
$$\times 10^{-7} + 5.1320024 \times 10^{-10}$$
$$= 0.5247542 \approx 0.52475$$

correct to 5 decimal places.

From the calculator

$\sin^{-1}(0.501) = 0.52475$ correct to 5 decimal places.

Worked Example 53

Prove the binomial expansion

$$(a+b)^n = a^n + na^{n-1}b + \frac{n(n-1)}{2!}a^{n-2}b^2$$
$$+ \frac{n(n-1)(n-2)}{3!}a^{n-3}b^3 + \ldots$$

using Taylor's theorem.

Solution 53

Let $f(a) = a^n$

$$f'(a) = na^{n-1}$$
$$f''(a) = n(n-1)a^{n-2}$$
$$f'''(a) = n(n-1)(n-2)a^{n-3}$$

$\ldots\ldots\ldots\ldots\ldots\ldots\ldots\ldots\ldots\ldots\ldots\ldots\ldots$

Using Taylor's theorem which states

$$f(x+h) = f(x) + hf'(x) + \frac{h^2}{2!}f''(x) + \frac{h^3}{3!}f'''(x) + \ldots$$

If $x = a$ and $h = b$

$$(a+b)^n = f(a) + bf'(a) + \frac{b^2}{2!}f''(a) + \frac{b^3}{3!}f'''(a) + \ldots$$

$$(a+b)^n = a^n + na^{n-1}b + \frac{n(n-1)}{2!}a^{n-2}b^2$$
$$+ \frac{n(n-1)(n-2)}{3!}a^{n-3}b^3 + \ldots$$

Successive Approximations

Consider the function $f(x) = \dfrac{1}{1-x}$

where x is less than unity.

Using long division, we have

$$
\begin{array}{r}
1 + x + x^2 \\
1-x\,\overline{\big)\,1} \\
\underline{1 - x} \\
x \\
\underline{x - x^2} \\
x^2 \\
\underline{x^2 - x^3} \\
x^3
\end{array}
$$

It is observed that

$$\dfrac{1}{1-x} = 1 + \dfrac{x}{1-x}$$

$$\dfrac{1}{1-x} = 1 + x + \dfrac{x^2}{1-x}$$

$$\dfrac{1}{1-x} = 1 + x + x^2 + \dfrac{x^3}{1-x}$$

Hence $1,\ 1+x,\ 1+x+x^2, \ldots$ are successive approximations to the function $f(x)$ and the respective errors are $\dfrac{x}{1-x},\ \dfrac{x^2}{1-x},\ \dfrac{x^3}{1-x}, \ldots$,

If $x = 0$, the successive approximations are all equal.

Therefore, the function can be written as $f(x) = a_0 + a_1 x + a_2 x^2 + a_3 x^3 + \ldots$

Maclaurin's (or Stirling's) Theorem (Proof)

Let $f(x) = a_0 + a_1 x + a_2 x^2 + a_3 x^3 + \ldots + a_n x^n + \ldots$

$f'(x) = a_1 + 2a_2 x + 3a_3 x^2 + \ldots + n a_n x^{n-1} + \ldots$

$f''(x) = 2a_2 + 3 \times 2a_3 x + \ldots$
$ + n(n-1) a_n x^{n-2} + \ldots$

$f'''(x) = 3 \times 2 \times 1 a_3 + \ldots$
$ + n(n-1)(n-2) a_n x^{n-3} + \ldots$

$f^{(n)}(x) = n(n-1)(n-2) \ldots 3.2.1\, a_n + \text{powers of } x.$

Putting $x = 0$ in each of the above equations, we have $f(0) = 0! a_0,\ f'(0) = 1! a_1,\ f''(0) = 2! a_2,\ f'''(0) = 3! a_3,$ and, in general, $f^{(n)}(0) = n! a_n$.

Thus $a_0 = f(0),\ a_1 = f'(0),\ a_2 = \dfrac{f''(0)}{2!},\ a_3 = \dfrac{f'''(0)}{3!},$

and, in general, $a_n = \dfrac{f^{(n)}(0)}{n!}$.

Therefore, the Maclaurin's Theorem is written as:

$$\boxed{\,f(x) = f(0)\dfrac{x^0}{0!} + f'(0)\dfrac{x^1}{1!} + f''(0)\dfrac{x^2}{2!} \\ + f'''(0)\dfrac{x^3}{3!} + \ldots + f^{(n)}(0)\dfrac{x^n}{n!} + \ldots\,}$$

This is a formula for expanding any function as a polynomial in ascending powers of the variable, assuming that such an expansion is possible.

WORKED EXAMPLE 54

Express the following functions:-

(i) $\cosh x$

(ii) $\sinh x$

(iii) $\cos x$

(iv) $\sin x$

(v) $\ln(1+x)$

as power series using Maclaurin's theorem as far as the term in x^4.

Solution 54

(i) $\quad f(x) = \cosh x \qquad f(0) = 1$
$ f'(x) = \sinh x \qquad f'(0) = 0$
$ f''(x) = \cosh x \qquad f''(0) = 1$
$ f'''(x) = \sinh x \qquad f'''(0) = 0$
$ f^{iv}(x) = \cosh x \qquad f^{iv}(0) = 1$

$$\boxed{\cosh x = 1 + \dfrac{x^2}{2!} + \dfrac{x^4}{4!}.}$$

(ii) $\quad f(x) = \sinh x \qquad f(0) = 0$
$ f'(x) = \cosh x \qquad f'(0) = 1$
$ f''(x) = \sinh x \qquad f''(0) = 0$
$ f'''(x) = \cosh x \qquad f'''(0) = 1$
$ f^{iv}(x) = \sinh x \qquad f^{iv}(0) = 0$

$$\boxed{\sinh x = x + \dfrac{x^3}{3!}.}$$

Maclaurin's Expansions — 61

(iii) $f(x) = \cos x \qquad f(0) = 1$

$f'(x) = -\sin x \qquad f'(0) = 0$

$f''(x) = -\cos x \qquad f''(0) = -1$

$f'''(x) = \sin x \qquad f'''(0) = 0$

$f^{iv}(x) = \cos x \qquad f^{iv}(0) = 1$

$$\boxed{\cos x = 1 - \frac{x^2}{2!} + \frac{x^4}{4!}.}$$

(iv) $f(x) = \sin x \qquad f(0) = 0$

$f'(x) = \cos x \qquad f'(0) = 1$

$f''(x) = -\sin x \qquad f''(0) = 0$

$f'''(x) = -\cos x \qquad f'''(0) = -1$

$f^{iv}(x) = \sin x \qquad f^{iv}(0) = 0$

$$\boxed{\sin x = x - \frac{x^3}{3!}.}$$

(v) $f(x) = \ln(1 + x) \qquad f(0) = 0$

$f'(x) = \dfrac{1}{1+x} \qquad f'(0) = 1$

$f''(x) = -\dfrac{1}{(1+x)^2} \qquad f''(0) = -1$

$f'''(x) = \dfrac{2}{(1+x)^3} \qquad f'''(0) = 2$

$f^{iv}(x) = \dfrac{-6}{(1+x)^4} \qquad f^{iv}(0) = -6$

$$\boxed{\ln(1+x) = x - \frac{x^2}{2} + \frac{x^3}{3} - \frac{x^4}{4}.}$$

WORKED EXAMPLE 55

Find the first five terms in the expansion of $\ln(1+\sin x)$, using Maclaurin's theorem.

Solution 55

$f(x) = \ln(1 + \sin x) \qquad f(0) = 0$

$f'(x) = \dfrac{\cos x}{1 + \sin x} \qquad f'(0) = 1$

$f''(x) = \dfrac{(-\sin x)(1 + \sin x) - \cos^2 x}{(1 + \sin x)^2}$

$= \dfrac{-\sin x - \sin^2 x - \cos^2 x}{(1 + \sin x)^2}$

$= -\dfrac{1 + \sin x}{(1 + \sin x)^2} = -\dfrac{1}{1 + \sin x}$

$f''(0) = -1$

$f'''(x) = \dfrac{\cos x}{(1 + \sin x)^2} \qquad f'''(0) = 1$

$f^{(iv)}(x) = \dfrac{(-\sin x)(1 + \sin x)^2 - 2\cos^2 x(1 + \sin x)}{(1 + \sin x)^4}$

$= \dfrac{(-\sin x)(1 + \sin x) - 2\cos^2 x}{(1 + \sin x)^3}$

$= \dfrac{-\sin x - \sin^2 x - 2\cos^2 x}{(1 + \sin x)^3} \Rightarrow f^{(iv)}(0) = -2$

$f^{(v)}(x) = \dfrac{\left\{\begin{array}{l}(-\cos x - 2\sin x \cos x + 4\cos x \sin x)(1 + \sin x)^3 \\ +(\sin x + \sin^2 x + 2\cos^2 x)3(1 + \sin x)^2 \cos x\end{array}\right\}}{(1 + \sin x)^6}$

$f^{(v)}(0) = \dfrac{-1 + 6}{1} = 5$

$$\boxed{\ln(1+\sin x) = x - \frac{x^2}{2} + \frac{x^3}{6} - \frac{x^4}{12} + \frac{x^5}{24}.}$$

WORKED EXAMPLE 56

If $\sin x = x - \dfrac{x^3}{3!} + \dfrac{x^5}{5!} - \dfrac{x^7}{7!} + \ldots$, find the expansion for $\cos x$.

Solution 56

$\sin x = x - \dfrac{x^3}{3!} + \dfrac{x^5}{5!} - \dfrac{x^7}{7!} + \ldots$

Differentiating with respect to x.

$$\frac{d}{dx}(\sin x) = 1 - \frac{3x^2}{3!} + \frac{5x^4}{5!} - \frac{7x^6}{7!} + \ldots$$

$$\boxed{\cos x = 1 - \frac{x^2}{2!} + \frac{x^4}{4!} - \frac{x^6}{6!} + \ldots}$$

WORKED EXAMPLE 57

Find the expansions of the following functions:-

(i) $\dfrac{1}{x^2 - 9}$

(ii) $\dfrac{1}{4 - x^2}$

(iii) $\dfrac{1}{\sqrt{5^2 - x^2}}$

using Maclaurin's theorem as far as the term in x^3.

Solution 57

(i) $f(x) = \dfrac{1}{x^2 - 9}$ $\quad f(0) = -\dfrac{1}{9}$

$f'(x) = -\dfrac{2x}{(x^2 - 9)^2}$ $\quad f'(0) = 0$

$f''(x) = -\dfrac{2(x^2 - 9)^2 - 2x \cdot 2(x^2 - 9) \cdot 2x}{(x^2 - 9)^4}$

$= -\dfrac{2(x^2 - 9) - 8x^2}{(x^2 - 9)^3} = \dfrac{18 + 6x^2}{(x^2 - 9)^3}$

$f''(0) = -\dfrac{18}{9^3} = -\dfrac{2}{81}$

$f'''(x) = \dfrac{\{-(4x + 16x)(x^2 - 9)^3 + [2(x^2 - 9) - 8x^2](3)(x^2 - 9)^2 (2x)\}}{(x^2 - 9)^6}$

$f'''(0) = 0$

$\dfrac{1}{x^2 - 9} = -\dfrac{1}{9} - \dfrac{2}{81} \dfrac{x^2}{2} = -\dfrac{1}{9} - \dfrac{x^2}{81}.$

$\boxed{\dfrac{1}{x^2 - 9} = -\dfrac{1}{9} - \dfrac{x^2}{81}}$

(ii) $f(x) = \dfrac{1}{4 - x^2}$ $\quad \boxed{f(0) = \dfrac{1}{4}}$

$f'(x) = -\dfrac{1 \times (-2x)}{(4 - x^2)^2}$

$= \dfrac{2x}{(4 - x^2)^2},$ $\quad \boxed{f'(0) = 0}$

$f''(x) = \dfrac{2(4 - x^2)^2 - 2x \cdot 2(4 - x^2)(-2x)}{(4 - x^2)^4}$

$= \dfrac{2(4 - x^2) + 8x^2}{(4 - x^2)^3}$

$= \dfrac{8 - 2x^2 + 8x^2}{(4 - x^2)^3} = \dfrac{8 + 6x^2}{(4 - x^2)^3},$

$f''(0) = \dfrac{8}{4^3} = \dfrac{1}{8}$ $\quad \boxed{f''(0) = \dfrac{1}{8}}$

$f'''(x) = \dfrac{12x(4 - x^2)^3 - (8 + 6x^2) \cdot 3(4 - x^2)^2(-2x)}{(4 - x^2)^6}$

$= \dfrac{12x(4 - x^2) + 6x(8 + 6x^2)}{(4 - x^2)^4}$

$\boxed{f'''(0) = 0}$

$\boxed{\dfrac{1}{4 - x^2} = \dfrac{1}{4} + \dfrac{1}{8}\dfrac{x^2}{2} = \dfrac{1}{4} + \dfrac{x^2}{16}.}$

(iii) $\dfrac{1}{\sqrt{5^2 - x^2}} = (25 - x^2)^{-\frac{1}{2}} = f(x)$ $\quad f(0) = \dfrac{1}{5}$

$f'(x) = -\dfrac{1}{2}(25 - x^2)^{-\frac{3}{2}}(-2x)$

$= x(25 - x^2)^{-\frac{3}{2}},$ $\quad \boxed{f'(0) = 0}$

$f''(x) = (25 - x^2)^{-\frac{3}{2}}$

$\quad -\dfrac{3}{2}x(25 - x^2)^{-\frac{5}{2}}(-2x)$

$= (25 - x^2)^{-\frac{3}{2}}$

$\quad + 3x^2(25 - x^2)^{-\frac{5}{2}}$ $\quad \boxed{f''(0) = \dfrac{1}{125}}$

$$f'''(x) = -\frac{3}{2}(25-x^2)^{-\frac{5}{2}}(-2x)$$
$$+ 6x(25-x^2)^{-\frac{5}{2}}$$
$$-\frac{15}{2}x^2(25-x^2)^{-\frac{7}{2}}(-2x)$$

$$f'''(0) = 0$$

$$\boxed{\frac{1}{\sqrt{5^2-x^2}} = \frac{1}{5} + \frac{1}{250}x^2.}$$

WORKED EXAMPLE 58

Using the expansion of $\frac{6}{x^2-9}$, derive the expansion of $\ln\frac{x-3}{x+3}$.

Solution 58

$$\frac{6}{x^2-9} = 6(x^2-9)^{-1}$$

$$= 6\left[x^2\left(1-\frac{9}{x^2}\right)\right]^{-1} = 6x^{-2}\left(1-\frac{9}{x^2}\right)^{-1}$$

$$= \frac{6}{x^2}\left[1 + (-1)\left(-\frac{9}{x^2}\right)\right.$$
$$\left. + (-1)(-2)\left(-\frac{9}{x^2}\right)^2\frac{1}{2!} + \ldots\right]$$

$$= \frac{6}{x^2} + \frac{54}{x^4} + \frac{486}{x^6} + \ldots$$

$$\int \frac{6}{x^2-9}dx = 6\left[\frac{1}{2\times 3}\ln\frac{x-3}{x+3}\right]$$

$$= \ln\frac{x-3}{x+3} \text{ when } |x| > 3$$

$$\ln\frac{x-3}{x+3} = \int\left(\frac{6}{x^2} + \frac{54}{x^4} + \frac{486}{x^6} + \ldots\right)dx$$

$$= \int\left(6x^{-2} + 54x^{-4} + 486x^{-6} + \ldots\right)dx$$

$$= \frac{6x^{-1}}{-1} + \frac{54x^{-3}}{-3} + \frac{486x^{-5}}{-5} + \ldots$$

$$= -\frac{6}{x} - \frac{18}{x^3} - \frac{486}{5x^5} + \ldots$$

Maclaurin's Expansions — 63

Leibnitz's Theorem

This theorem is a generalisation of the product rule in differentiation.

The nth derivative of the product is given by

$$(uv)_n = u_n v + {}^nC_1 u_{n-1}v_1 + {}^nC_2 u_{n-2}v_2 + \ldots$$
$$+ {}^nC_r u_{n-r}v_r + \ldots + uv_n.$$

The product rule $\frac{d}{dx}(uv) = \frac{du}{dx}v + u\frac{dv}{dx}$ which can be written as $(uv)_1 = u_1 v + uv_1$

$$\frac{d^2}{dx^2}(uv) = \frac{d}{dx}\left(\frac{du}{dx}v + u\frac{dv}{dx}\right)$$
$$= \frac{d^2u}{dx^2}v + \frac{du}{dx}\frac{dv}{dx} + \frac{du}{dx}\frac{dv}{dx} + u\frac{d^2v}{dx^2}$$
$$= u_2 v + 2u_1 v_1 + uv_2$$
$$= u_2 v + {}^2C_1 u_1 v_1 + uv_2$$

$$\frac{d^3}{dx^3}(uv) = \frac{d}{dx}\left(\frac{d^2}{dx^2}(uv)\right)$$
$$= \frac{d}{dx}\left(u_2 v + {}^2C_1 u_1 v_1 + uv_2\right)$$
$$= u_3 v + u_2 v_1 + {}^2C_1 u_2 v_1$$
$$\quad + {}^2C_1 u_1 v_2 + u_1 v_2 + uv_3$$
$$= u_3 v + 3u_2 v_1 + 3u_1 v_2 + uv_3$$
$$= u_3 v + {}^3C_1 u_2 v_1 + {}^3C_2 u_1 v_2 + uv_3$$

$$\boxed{\frac{d^n}{dx^n}(uv) = (uv)_n = u_n v + {}^nC_1 u_{n-1}v_1 + {}^nC_2 u_{n-2}v_2 \\ + \ldots + {}^nC_r u_{n-r}v_r + \ldots + uv_n.}$$

Numerical Methods for the Solution of Differential Equations
Polynomial Approximations Using Taylor Series

The Taylor expansion of $f(x)$ about $x - x_0$ is given by

$$y = y_0 + \left(\frac{dy}{dx}\right)_0 (x-x_0)$$
$$+ \left(\frac{d^2y}{dx^2}\right)_0 \frac{(x-x_0)^2}{2!} + \left(\frac{d^3y}{dx^3}\right)_0 \frac{(x-x_0)^3}{3!} + \ldots$$

where $y_0, \left(\frac{dy}{dx}\right)_0, \left(\frac{d^2y}{dx^2}\right)_0, \left(\frac{d^3y}{dx^3}\right)_0, \ldots$ are the values at $x = x_0$

WORKED EXAMPLE 59

Find, as a series of ascending powers of x up to and including the term in x^5, an approximate solution to the differential equation $\dfrac{dy}{dx} = ye^{-x}$, where $y = 1$ when $x = 0$.

Solution 59

$$\boxed{\dfrac{dy}{dx} = ye^{-x}} \quad \ldots(1)$$

differentiating with respect to x

$$\boxed{\dfrac{d^2y}{dx^2} = \dfrac{dy}{dx}e^{-x} - ye^{-x}} \quad \ldots(2)$$

differentiating with respect to x

$$\dfrac{d^3y}{dx^3} = \dfrac{d^2y}{dx^2}e^{-x} - \dfrac{dy}{dx}e^{-x} - \dfrac{dy}{dx}e^{-x} + ye^{-x}$$

$$\boxed{\dfrac{d^3y}{dx^3} = \dfrac{d^2y}{dx^2}e^{-x} - 2\dfrac{dy}{dx}e^{-x} + ye^{-x}} \quad \ldots(3)$$

differentiating with respect to x

$$\dfrac{d^4y}{dx^4} = \dfrac{d^3y}{dx^3}e^{-x} - \dfrac{d^2y}{dx^2}e^{-x} - 2\dfrac{d^2y}{dx^2}e^{-x}$$
$$+ 2\dfrac{dy}{dx}e^{-x} + \dfrac{dy}{dx}e^{-x} - ye^{-x}$$

$$\boxed{\dfrac{d^4y}{dx^4} = \dfrac{d^3y}{dx^3}e^{-x} - 3\dfrac{d^2y}{dx^2}e^{-x} + 3\dfrac{dy}{dx}e^{-x} - ye^{-x}} \quad \ldots(4)$$

differentiating with respect to x

$$\dfrac{d^5y}{dx^5} = \dfrac{d^4y}{dx^4}e^{-x} - \dfrac{d^3y}{dx^3}e^{-x} - 3\dfrac{d^3y}{dx^3}e^{-x} + 3\dfrac{d^2y}{dx^2}e^{-x}$$
$$+ 3\dfrac{d^2y}{dx^2}e^{-x} - 3\dfrac{dy}{dx}e^{-x} - \dfrac{dy}{dx}e^{-x} + ye^{-x}$$

$$\boxed{\dfrac{d^5y}{dx^5} = \dfrac{d^4y}{dx^4}e^{-x} - 4\dfrac{d^3y}{dx^3}e^{-x} + 6\dfrac{d^2y}{dx^2}e^{-x} - 4\dfrac{dy}{dx}e^{-x} + ye^{-x}} \quad \ldots(5)$$

$y = 1$ when $x = 0$ in (1)

$$\left(\dfrac{dy}{dx}\right)_0 = y_0 e^{-x_0}, \left(\dfrac{dy}{dx}\right)_0 = 1e^0 = 1 \quad \boxed{\left(\dfrac{dy}{dx}\right)_0 = 1}$$

$y = 1$ when $x = 0$ in (2)

$$\left(\dfrac{d^2y}{dx^2}\right)_0 = \left(\dfrac{dy}{dx}\right)_0 e^{-x_0} - y_0 e^{-x_0},$$

$$\left(\dfrac{d^2y}{dx^2}\right)_0 = (1)(1) - (1)(1) = 0 \quad \boxed{\left(\dfrac{d^2y}{dx^2}\right)_0 = 0}$$

$y = 1$ when $x = 0$, $\left(\dfrac{dy}{dx}\right)_0 = 1$, $\left(\dfrac{d^2y}{dx^2}\right)_0 = 0$ in (3)

$$\left(\dfrac{d^3y}{dx^3}\right)_0 = \left(\dfrac{d^2y}{dx^2}\right)_0 e^{-x_0} - 2\left(\dfrac{dy}{dx}\right)_0 e^{-x_0} + y_0 e^{-x_0}$$

$$\left(\dfrac{d^3y}{dx^3}\right)_0 = (0)(1) - 2(1)(1) + (1)(1) = -1,$$

$$\boxed{\left(\dfrac{d^3y}{dx^3}\right)_0 = -1}$$

$y = 1$ when $x = 0$, $\left(\dfrac{dy}{dx}\right)_0 = 1,$

$\left(\dfrac{d^2y}{dx^2}\right)_0 = 0, \left(\dfrac{d^3y}{dx^3}\right)_0 = -1$ in (4)

$$\left(\dfrac{d^4y}{dx^4}\right)_0 = \left(\dfrac{d^3y}{dx^3}\right)_0 e^{-x_0} - 3\left(\dfrac{d^2y}{dx^2}\right)_0 e^{-x_0}$$
$$+ 3\left(\dfrac{dy}{dx}\right)_0 e^{-x_0} - y_0 e^{-x_0}$$

$$\left(\dfrac{d^4y}{dx^4}\right)_0 = (-1)(1) - 3(0)(1) + 3(1)(1) - (1)(1) = 1$$

$$\boxed{\left(\dfrac{d^4y}{dx^4}\right)_0 = 1}$$

$y = 1$ when $x = 0$, $\left(\dfrac{dy}{dx}\right)_0 = 1, \left(\dfrac{d^2y}{dx^2}\right)_0 = 0,$

$\left(\dfrac{d^3y}{dx^3}\right)_0 = -1, \left(\dfrac{d^4y}{dx^4}\right)_0 = 1$ in (5)

$$\left(\dfrac{d^5y}{dx^5}\right)_0 = \left(\dfrac{d^4y}{dx^4}\right)_0 e^{-x_0} - 4\left(\dfrac{d^3y}{dx^3}\right)_0 e^{-x_0}$$
$$+ 6\left(\dfrac{d^2y}{dx^2}\right)_0 e^{-x_0} - 4\left(\dfrac{dy}{dx}\right)_0 e^{-x_0} + y_0 e^{-x_0}$$

Maclaurin's Expansions

$$\left(\frac{d^5 y}{dx^5}\right)_0 = (1)(1) - 4(-1)(1) + 6(0)(1)$$

$$- 4(1)(1) + (1)(1) = 2 \quad \boxed{\left(\frac{d^5 y}{dx^5}\right)_0 = 2}$$

Applying Taylor's expansion of $f(x)$ about $x - x_0$

$$y \approx y_0 + \left(\frac{dy}{dx}\right)_0 (x - x_0) + \left(\frac{d^2 y}{dx^2}\right)_0 \frac{(x - x_0)^2}{2!}$$

$$+ \left(\frac{d^3 y}{dx^3}\right)_0 \frac{(x - x_0)^3}{3!} + \cdots$$

$$y \approx 1 + (1)x + (0)\frac{x^2}{2!} + (-1)\frac{x^3}{3!} + (1)\frac{x^4}{4!} + (2)\frac{x^5}{5!}$$

$$y \approx 1 + x - \frac{1}{6}x^3 + \frac{1}{24}x^4 + \frac{1}{60}x^5$$

The exact solution of the differential equation $\frac{dy}{dx} = ye^{-x}$, where $y = 1$ when $x = 0$.

$\frac{dy}{y} = e^{-x}dx$, separable variables integrating both sides.

$$\int \frac{dy}{y} = \int e^{-x} dx$$

$\ln y = -e^{-x} + c$, $\ln 1 = -e^0 + c$, $\boxed{c = 1}$,

$\ln y = -e^{-x} + 1$

$$\boxed{y = e^{-e^{-x}+1}}$$

Approximate values   Exact value

$$y \approx 1 + x - \frac{1}{6}x^3 + \frac{1}{24}x^4 + \frac{1}{60}x^5 \qquad y = e^{-e^{-x}+1}$$

x		x	
0	$y \approx 1$	0	$y = 1$
0.1	$y \approx 1.0998377$	0.1	$y = 1.0998377$
0.5	$y \approx 1.4822917$	0.5	$y = 1.481138$
1	$y \approx 1.8916667$	1.0	$y = 1.8815964$
1.5	$y \approx 2.275$	1.5	$y = 2.1746546$
2.0	$y \approx 2.866667$	2.0	$y = 2.3742099$

For values below $x = 1$ the results are approximately correct by the two methods. For values greater than 1 the values converge very slowly and hence the errors are very large.

WORKED EXAMPLE 60

Find, as a series of ascending powers of x up to and including the term in x^4, an approximate solution to the differential equation $\frac{dy}{dx} = \frac{y^2 - 1}{x}$, where $y = 2$ when $x = 1$.

Solution 60

$$\boxed{x\frac{dy}{dx} = y^2 - 1} \qquad \ldots (1)$$

$$x_0 \left(\frac{dy}{dx}\right)_0 = y_0^2 - 1, \quad (1)\left(\frac{dy}{dx}\right)_0 = (2)^2 - 1$$

$$\boxed{\left(\frac{dy}{dx}\right)_0 = 3.}$$

Differentiating with respect to x equation (1)

$$\boxed{\frac{dy}{dx} + x\frac{d^2 y}{dx^2} = 2y\frac{dy}{dx}} \qquad \ldots (2)$$

repeating the differentiation

$$\frac{d^2 y}{dx^2} + \frac{d^2 y}{dx^2} + x\frac{d^3 y}{dx^3} = 2\frac{dy}{dx}\frac{dy}{dx} + 2y\frac{d^2 y}{dx^2}$$

$$\boxed{2\frac{d^2 y}{dx^2} + x\frac{d^3 y}{dx^3} = 2\left(\frac{dy}{dx}\right)^2 + 2y\frac{d^2 y}{dx^2}} \qquad \ldots (3)$$

Differentiating with respect to x equation (3)

$$2\frac{d^3 y}{dx^3} + \frac{d^3 y}{dx^3} + x\frac{d^4 y}{dx^4}$$

$$= 4\left(\frac{dy}{dx}\right)\frac{d^2 y}{dx^2} + 2\frac{dy}{dx}\frac{d^2 y}{dx^2} + 2y\frac{d^3 y}{dx^3}$$

$$\boxed{3\frac{d^3 y}{dx^3} + x\frac{d^4 y}{dx^4} = 6\frac{dy}{dx}\frac{d^2 y}{dx^2} + 2y\frac{d^3 y}{dx^3}} \qquad \ldots (4)$$

From (2)

$$\left(\frac{dy}{dx}\right)_0 + x_0\left(\frac{d^2 y}{dx^2}\right)_0 = 2y_0\left(\frac{dy}{dx}\right)_0, \; 3 + (1)\left(\frac{d^2 y}{dx^2}\right)_0$$

$$= 2(2)(3)$$

$$\boxed{\left(\frac{d^2 y}{dx^2}\right)_0 = 9},$$

From (3)

$$2\left(\frac{d^2y}{dx^2}\right)_0 + x_0\left(\frac{d^3y}{dx^3}\right)_0 = 2\left(\frac{dy}{dx}\right)_0^2 + 2y_0\left(\frac{d^2y}{dx^2}\right)_0$$

$$2(9) + (1)\left(\frac{d^3y}{dx^3}\right)_0 = 2(3)^2 + 2(2)(9) \quad \boxed{\left(\frac{d^3y}{dx^3}\right)_0 = 36}$$

From (4) $3\left(\frac{d^3y}{dx^3}\right)_0 + x_0\left(\frac{d^4y}{dx^4}\right)_0$

$$= 6\left(\frac{dy}{dx}\right)_0\left(\frac{d^2y}{dx^2}\right)_0 + 2y_0\left(\frac{d^3y}{dx^3}\right)_0$$

$$3(36) + (1)\left(\frac{d^4y}{dx^4}\right)_0 = 6(3)(9) + 2(2)(36)$$

$$\left(\frac{d^4y}{dx^4}\right)_0 = 162 + 144 - 108 = 198$$

$$y \approx y_0 + \left(\frac{dy}{dx}\right)_0\frac{(x-x_0)}{1!} + \left(\frac{d^2y}{dx^2}\right)_0\frac{(x-x_0)^2}{2!}$$

$$+ \left(\frac{d^3y}{dx^3}\right)_0\frac{(x-x_0)^3}{3!} + \ldots, \text{ when } x_0 = 1$$

$$y \approx 2 + \frac{3(x-1)}{1!} + \frac{9(x-1)^2}{2!}$$

$$+ \frac{36(x-1)^3}{3!} + \ldots; \text{ when } x = 1, y = 2,$$

when $x = 0.5$, $y \approx 2 + 3(-0.5) + \frac{9(-0.5)^2}{2} + \frac{36(-0.5)^3}{3!} + \frac{198(-0.5)^4}{4!}$

$y \approx 2 - 1.5 + 1.125 - 0.75 + 0.515625 - \ldots = 1.390625$
We have to take many more terms to approximate to the exact value of 1.1818182.

The exact solution of the differential equation

$$\frac{dy}{y^2-1} = \frac{dx}{x}$$

$$\int \frac{dy}{(y-1)(y+1)} = \int \frac{dx}{x}$$

$$\frac{1}{(y-1)(y+1)} \equiv \frac{A}{y-1} + \frac{B}{y+1}$$

$$1 \equiv A(y+1) + B(y-1)$$

If $y = 1$, $A = \frac{1}{2}$; if $y = -1$, $B = -\frac{1}{2}$.

$$\frac{1}{(y-1)(y+1)} \equiv \frac{\frac{1}{2}}{y-1} - \frac{\frac{1}{2}}{y+1}$$

$$\int\left[\frac{\frac{1}{2}}{(y-1)} - \frac{\frac{1}{2}}{(y+1)}\right]dy = \frac{1}{2}\ln|y-1| - \frac{1}{2}\ln|y+1|$$

$$\frac{1}{2}\ln\frac{y-1}{y+1} = \ln x + \ln A \quad \text{where } x > 0 \quad y > 1$$

when $y = 2$, $x = 1$, $\frac{1}{2}\ln\frac{1}{3} = \ln 1 + \ln A$

$$\frac{1}{2}\ln\frac{y-1}{y+1} = \ln x + \frac{1}{2}\ln\frac{1}{3}$$

$$\ln\sqrt{\frac{y-1}{y+1}} = \ln x\sqrt{\frac{1}{3}} \qquad \frac{y-1}{y+1} = \frac{1}{3}x^2$$

$$3y - 3 = yx^2 + x^2 \quad 3y - yx^2 = x^2 + 3 \quad y = \frac{x^2+3}{3-x^2}$$

x	Approximate values	Exact values
1	2	2
0.5	1.3180416	1.1818182

WORKED EXAMPLE 61

Find the solution, in ascending powers of x up to and including the term x^3, of the differential equation $\frac{d^2y}{dx^2} + (x+1)\frac{dy}{dx} + y = 0$ given that, when $x = 0$, $y = 1$ and $\frac{dy}{dx} = 2$.

Solution 61

$$\frac{d^2y}{dx^2} + (x+1)\frac{dy}{dx} + y = 0 \qquad \ldots(1)$$

$$\left(\frac{d^2y}{dx^2}\right)_0 + (x_0+1)\left(\frac{dy}{dx}\right)_0 + y_0 = 0$$

$$x_0 = 0, y_0 = 1, \left(\frac{dy}{dx}\right)_0 = 2$$

$$\left(\frac{d^2y}{dx^2}\right)_0 + (2) + 1 = 0 \qquad \left(\frac{d^2y}{dx^2}\right)_0 = -3$$

Differentiating equation (1) with respect to x

$$\frac{d^3y}{dx^3} + \frac{dy}{dx} + (x+1)\frac{d^2y}{dx^2} + \frac{dy}{dx} = 0$$

$$\left(\frac{d^3y}{dx^3}\right)_0 + \left(\frac{dy}{dx}\right)_0 + (x_0+1)\left(\frac{d^2y}{dx^2}\right)_0 + \left(\frac{dy}{dx}\right)_0 = 0$$

$$\left(\frac{d^3y}{dx^3}\right)_0 + 2 + (0+1)(-3) + (2) = 0 \quad \left(\frac{d^3y}{dx^3}\right)_0 = -1$$

$$y \approx y_0 + \left(\frac{dy}{dx}\right)_0 (x-x_0) + \left(\frac{d^2y}{dx^2}\right)_0 \frac{(x-x_0)^2}{2!} + \ldots$$

$$y \approx 1 + 2x + (-3)\frac{x^2}{2} + (-1)\frac{x^3}{6}$$

$$y \approx 1 + 2x - \frac{3}{2}x^2 - \frac{1}{6}x^3$$

Exercises 11

1. If $\cos x = 1 - \frac{x^2}{2!} + \frac{x^4}{4!} - \frac{x^6}{6!} + \ldots$, find the power series for $\sin x$.

2. Obtain the first three terms of the expansion $\sin^{-1} x$ using the Maclaurin's theorem.

3. If $\tan^{-1} x = a_0 + a_1 x + a_2 x^2 + a_3 x^3 + a_4 x^4$, determine the coefficients a_0, a_1, a_2, a_3, a_4.

4. Determine the expansion for $\frac{1}{1+x^2}$ using:-
   (a) the binomial theorem
   (b) the Maclaurin's theorem.

5. Using the expansion of $\frac{1}{25-x^2}$, derive the expansion of $\ln \frac{5+x}{5-x}$, $|x| < 5$.

6. Using the expansion of $\frac{1}{(x^2+4)^{\frac{1}{2}}}$, derive the expansion of $\sinh^{-1}\frac{x}{2}$.

7. Using the expansion of $\frac{1}{\sqrt{1-x^2}}$, derive the expansion of $\sin^{-1} x$ ($|x| < 1$).

8. Use Maclaurin's theorem to obtain the following expansions:-
   (i) $\sinh^2 x$
   (ii) $\log_e (1-x^2)$
   (iii) $\sin^2 x$
   (iv) $\cos^2 x$.

9. Use Taylor's theorem to obtain approximate values of the following:-
   (i) $\tan 45°2'$
   (ii) $\log_e 1.001$.

10. Determine the expansion of $\log_e (1 + x + x^2)$. Hence determine $\ln 1.0204$.

    Hint: use $1 - x^3 = (1-x)(1+x+x^2)$ provided that $-1 \leq x < 1$.

11. Determine the expansions:-
    (i) $\log_e (1 - 4x)$
    (ii) $\log_e (3 + 5x)$
    (iii) $\log_e (2 + 7x)$.

    State in each case the range of values of x for which the expansions are valid.

12. Use the expansion of $\tan^{-1} x$ to deduce the result

    $$\pi = 4\left(1 - \frac{1}{3} + \frac{1}{5} - \frac{1}{7} + \ldots\right).$$

13. Use the expansion of $\sin^{-1} x$ to deduce the result

    $$\pi = 6\left(\frac{1}{2} + \frac{1}{2 \times 3 \times 2^3} + \frac{1 \times 3}{2 \times 4 \times 5 \times 2^5} + \ldots\right).$$

14. Find the first four terms in the expansion of $\cos^{-1} x$ and hence determine the value of π to three significant figures. You may use the expansion of $\cos^{-1} x$ or derive it by Maclaurin's theorem.

15. By means of the Taylor's series method, derive the solution, as a power series of ascending powers of x as far as the term in x^6, of the differential equation

    $$\frac{d^2y}{dx^2} - 5x\frac{dy}{dx} + 2y = 0, \text{ given that } y = 2, \frac{dy}{dx} = 1$$
    when $x = 0$.

    $$\left(\text{Ans. } 2 + x - 2x^2 - 2x^3 - \frac{13}{6}x^4 - \frac{13}{10}x^5 - \frac{13}{10}x^6\right)$$

16. Use Maclaurin's theorem to obtain the first three non-zero terms of the series expansion for $\sinh x$. Differentiate this to obtain the series for $\cosh x$.

Hence or otherwise, obtain series expansions in terms of x up to and including the term in x^3 for:-

(i) $e^{-\cosh x}$

(ii) $e^{-\sinh x}$

Ans. (i) $\dfrac{1}{3} - x^2$

(ii) $1 - x + \dfrac{x^2}{2} - \dfrac{x^3}{3}$

17. If $y = \tan x$, find the first six derivatives using the notation $y_r = \dfrac{d^r y}{dx^r}$

18. If $y = \ln \sin x$, find the first four derivatives. Obtain the Maclaurin's expansion of y in terms of x up to and including the term in x^5.

Ans. $y_1 = 1 + y^2$, $y_2 = 2yy_1$, $y_3 = 2y_1^2 + 2y_1y_2$, $y_4 = 6y_1y_2 + 2yy_3$, $y_5 = 6y_2^2 + 8y_1y_3 + 2yy_4$

19. State the expansion of $y = \sinh nx$ as a series in ascending powers of x up to and including the term in n^5.

Ans. $nx + \dfrac{n^3 x^3}{3!} + \dfrac{n^5 x^5}{5!}$

12
Additional Examples

C1
Example 1
Differentiate the following functions
(a) $y = 5x + 3x^2 - \dfrac{1}{x^3}$, $x \neq 0$

(b) $y = \dfrac{(5 + \sqrt{x})^2}{\sqrt{x}}$, $x \neq 0$

(c) $y = (3x - 1)(x + 5)$

(d) $y = \dfrac{x^2 - 3x + 2}{x - 1}$

Solution 1
(a) $y = 5x + 3x^2 - \dfrac{1}{x^3}$

$y = 5x + 3x^2 - x^{-3}$

$\dfrac{dy}{dx} = 5 + 6x + 3x^{-4}$

$= 5 + 6x + \dfrac{3}{x^4}$

(b) $y = \dfrac{(5 + \sqrt{x})^2}{\sqrt{x}}$

$y = \dfrac{25 + 10\sqrt{x} + x}{\sqrt{x}}$

$y = 25x^{-\frac{1}{2}} + 10 + x^{\frac{1}{2}}$

$\dfrac{dy}{dx} = -\dfrac{25}{2}x^{-\frac{3}{2}} + \dfrac{1}{2}x^{-\frac{1}{2}}$

$\dfrac{dy}{dx} = -\dfrac{12.5}{\sqrt{x^3}} + \dfrac{1}{2\sqrt{x}}$

(c) $y = (3x - 1)(x + 5)$

$y = 3x^2 - x + 15x - 5$

$y = 3x^2 + 14x - 5$

$\dfrac{dy}{dx} = 6x + 14$

(d) $y = \dfrac{x^2 - 3x + 2}{x - 1}$

$y = \dfrac{(x - 1)(x - 2)}{x - 1}$

$y = x - 2$

$\dfrac{dy}{dx} = 1$

Example 2
The curve C has equation

$y = x^3 - 4x^2 - 4x + 16$.

The point P has coordinates $(2, 0)$

(a) Show that P lies on C.

(b) Find the equation of the tangent to C at P, giving your answer in the form $y = mx + c$, where m and c are constants.

Another point Q also lies on C. The tangent to C at Q is parallel to the tangent to C at P.

(c) Find the coordinates of Q.

Solution 2
(a) $f(x) = x^3 - 4x^2 - 4x + 16$

$f(2) = 8 - 16 - 8 + 16 = 0$

$\therefore$ P lies on C

(b) $f'(x) = 3x^2 - 8x - 4$

the gradient at any point on the curve.
When $x = 2$, $f'(2) = 12 - 16 - 4$

$m = -8$

$y = mx + c$

$y = -8x + c$

At $P(2, 0)$

$0 = -8(2) + c \Rightarrow c = 16$

$\therefore \quad y = -8x + 16$ the equation of tangent at P

(c) $3x^2 - 8x - 4 = -8$

$3x^2 - 8x + 4 = 0$

$x = \dfrac{8 \pm \sqrt{64 - 48}}{6}, \quad x = 2 \quad \text{or} \quad x = \dfrac{2}{3}$

$f\left(\dfrac{2}{3}\right) = \left(\dfrac{2}{3}\right)^3 - 4\left(\dfrac{2}{3}\right)^2 - 4\left(\dfrac{2}{3}\right) + 16$

$= \dfrac{8}{27} - \dfrac{16}{9} - \dfrac{8}{3} + 16$

$= \dfrac{8 - 48 - 72 + 16 \times 27}{27}$

$= \dfrac{320}{27}$

$\therefore \quad Q\left(\dfrac{2}{3}, \dfrac{320}{27}\right).$

C2
Example 3

The curve C has equation.

$y = (x - 1)(x^2 - 9)$ ... (1)

(a) Determine the coordinates of the intersections with the axes.

(b) Show that the derivative of (1) is given by
$\dfrac{dy}{dx} = 3x^2 - 2x - 9$

(c) Find the exact x-coordinates of the turning points.

(d) Sketch the curve and label the coordinates.

(e) Find $\dfrac{d^2y}{dx^2}$ and hence determine the nature of the turning points of C.

Solution 3

(a) $f(x) = (x - 1)(x^2 - 9) = x^3 - x^2 - 9x + 9$

$f(0) = (-1)(-9) = 9, \quad P(0, 9)$
the intersection of C with the y-axis.

$f(x) = 0, \quad x = 1, \quad x = \pm 3$

$Q(1, 0), R(3, 0), S(-3, 0)$
the intersections of C with the x-axis

(b) $f'(x) = \dfrac{dy}{dx} = 3x^2 - 2x - 9$

(c) $3x^2 - 2x - 9 = 0$ for turning points

$x = \dfrac{2 \pm \sqrt{4 + 108}}{6} = \dfrac{2 \pm 2\sqrt{28}}{6} = \dfrac{1 \pm \sqrt{28}}{3}$

$x = \dfrac{1 \pm 2\sqrt{7}}{3}, \quad x = \dfrac{1}{3} + \dfrac{2}{3}\sqrt{7}, \text{ or } x = \dfrac{1}{3} - \dfrac{2}{3}\sqrt{7}$

(d)

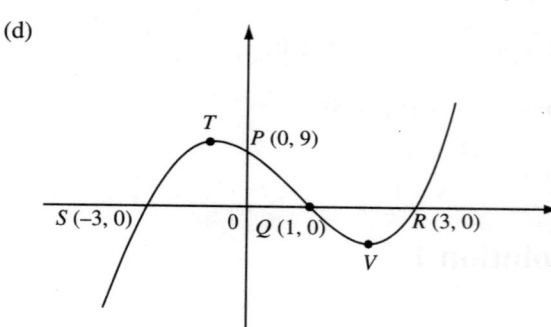

$T\left(\dfrac{1}{3} - \dfrac{2}{3}\sqrt{7}, y_1\right) V\left(\dfrac{1}{3} + \dfrac{2}{3}\sqrt{7}, y_2\right)$

(e) $\dfrac{d^2y}{dx^2} = 6x - 2$

when $x = \dfrac{1 + 2\sqrt{7}}{3} = 2.1$ to 1 d.p.

$\dfrac{d^2y}{dx^2} = 6x - 2 = 10.6$ to 1 d.p.

$\therefore$ minimum

when $x = \dfrac{1 - 2\sqrt{7}}{3} = -1.43$ to 2 d.p.

$\dfrac{d^2y}{dx^2} = -10.6$ to 2 d.p.

$\therefore$ maximum

Example 4

The curve C has equation

$$f(x) = \frac{2}{3}x^3 - \frac{7}{2}x^2 + 3x + 2.$$

(a) Find $f'(x)$.

(b) Find the coordinates of the turning points of C.

(c) Find $f''(x)$.

(d) Hence determine the nature of the turning points of C.

(e) Sketch the curve, and insert the coordinates of the turning points.

Solution 4

(a) $f(x) = \frac{2}{3}x^3 - \frac{7}{2}x^2 + 3x + 2$

$f'(x) = 2x^2 - 7x + 3$

(b) $f'(x) = 0$ for turning points
$2x^2 - 7x + 3 = (2x - 1)(x - 3) = 0$
$x = \frac{1}{2}$ or $x = 3$

(c) $f''(x) = 4x - 7$

(d) $f''\left(\frac{1}{2}\right) = 4\left(\frac{1}{2}\right) - 7 = -5 < 0 \therefore$ maximum

$f''(3) = 4(3) - 7 = 5 > 0 \therefore$ minimum

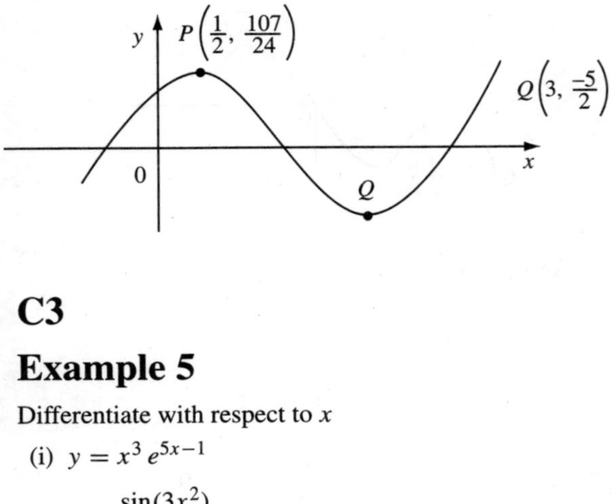

[Graph showing curve with maximum $P\left(\frac{1}{2}, \frac{107}{24}\right)$ and minimum $Q\left(3, \frac{-5}{2}\right)$]

C3
Example 5

Differentiate with respect to x

(i) $y = x^3 e^{5x-1}$

(ii) $y = \frac{\sin(3x^2)}{e^{4x}}$

(iii) $y = 5\cos(3x - 5)$

(iv) $y = 5\ln 3x - \sin 5x + e^{3x}$

(v) $y = (x^3 - 3)^{\frac{5}{2}}$

Solution 5

(i) $y = x^3 e^{5x-1}$

$\frac{dy}{dx} = 3x^2 e^{5x-1} + x^3 \times 5 e^{5x-1}$

$\frac{dy}{dx} = 3x^2 e^{5x-1} + 5x^3 e^{5x-1}$

(ii) $y = \frac{\sin(3x^2)}{e^{4x}}$

$\frac{dy}{dx} = \frac{6x\cos(3x^2)e^{4x} - \sin(3x^2)4e^{4x}}{(e^{4x})^2}$

$\frac{dy}{dx} = \frac{6xe^{4x}\cos(3x^2) - 4e^{4x}\sin(3x^2)}{e^{8x}}$

$\frac{dy}{dx} = \frac{6x\cos(3x^2) - 4\sin(3x^2)}{e^{4x}}$

(iii) $y = 5\cos(3x - 5)$

$\frac{dy}{dx} = -5 \times 3\sin(3x - 5)$

$\frac{dy}{dx} = -15\sin(3x - 5)$

(iv) $y = 5\ln 3x - \sin 5x + e^{3x}$

$\frac{dy}{dx} = \frac{5}{x} - 5\cos 5x + 3e^{3x}$

(v) $y = (x^3 - 3)^{\frac{5}{2}}$

$\frac{dy}{dx} = \frac{5}{2}(x^3 - 3)^{\frac{3}{2}}(3x^2)$

$= \frac{15x^2}{2}(x^3 - 3)^{\frac{3}{2}}.$

Example 6

A heated metal ball is dropped into a liquid. As the ball cools, its temperature, $T°C$, t minutes after it enters the liquid is given by

$$T = 500\,e^{-0.01t} + 50, t \geq 0.$$

(a) Find the temperature of the ball as it enters the liquid.
(b) Find the value of t for which $T = 250$, giving your answer to 3 significant figures.
(c) Find the rate at which the temperature of the ball is decreasing at the instant when $t = 75$.

Solution 6

(a) $T = 500 e^{-0.01t} + 50$

when $t = 0$

$T = 500 + 50 = 550°C$

(b) $300 = 500 e^{-0.01t} + 50$

$300 - 50 = 500 e^{-0.01t}$

$\dfrac{250}{500} = e^{-0.01t}$

$2 = e^{0.01t}$

Taking logs to the base e

$\ln 2 = 0.01t$

$t = \dfrac{\ln 2}{0.01} = 100 \ln 2$

$= 69.3$ to 3 s.f.

$\dfrac{dT}{dt} = -500 \times 0.01 \, e^{-0.01t}$

$= -5 e^{-0.01t}$

$= -5 e^{-0.01(75)}$

$= -2.36°C/\min$ to 3 s.f.

Example 7

(a) The curve $y = \dfrac{x}{x^2+1}$ has two turning points, find the coordinates and determine their nature.
(b) Sketch the curve, showing that $y \to 0$ as $x \to \infty$ or $x \to -\infty$

Solution 7

(a) $\dfrac{dy}{dx} = \dfrac{1(x^2+1) - x(2x)}{(x^2+1)^2}$

$= \dfrac{x^2 + 1 - 2x^2}{(x^2+1)^2} = \dfrac{-x^2+1}{(x^2+1)^2}$

$\dfrac{dy}{dx} = 0$ for turning points

$-x^2 + 1 = 0$

$x^2 = 1 \Rightarrow x = \pm 1$

$\dfrac{d^2y}{dx^2} = \dfrac{-2x(x^2+1)^2 - (-x^2+1)2(x^2+1)2x}{(x^2+1)^4}$

$= \dfrac{-2x(x^2+1) - 4x(-x^2+1)}{(x^2+1)^3}$

$= \dfrac{-2x^3 - 2x + 4x^3 - 4x}{(x^2+1)^3}$

$= \dfrac{2x^3 - 6x}{(x^2+1)^3} = \dfrac{2x(x^2-3)}{(x^2+1)^3}$

when $x = 1$, $\dfrac{d^2y}{dx^2} = \dfrac{2(1)(-2)}{8} = -\dfrac{1}{2} < 0$

∴ maximum

when $x = -1$, $\dfrac{d^2y}{dx^2} = \dfrac{2(-1)(-2)}{8} = \dfrac{1}{2} > 0$

∴ minimum

The coordinates are $\left(1, \dfrac{1}{2}\right)$ and $\left(-1, -\dfrac{1}{2}\right)$.

(b) $f(x) = \dfrac{x}{x^2+1}$

$f(0) = 0$

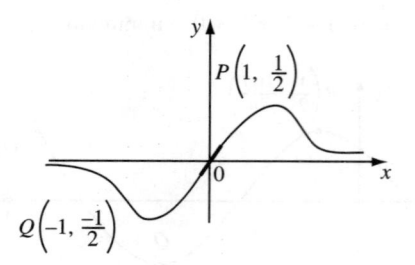

As $x \to \infty$, $y = \dfrac{x}{x^2\left(1 + \dfrac{1}{x^2}\right)}$

$= \dfrac{1}{x\left(1 + \dfrac{1}{x^2}\right)}$

$= \dfrac{1}{x(1+0)}, y \to 0$

as $x \to -\infty$, $y \to 0$.

C4
Example 8

(a) Sketch the curve represented by the parametric equations
$x = 3t^2$ and $y = 3t - 3t^3$

(b) Determine $\frac{dx}{dt}$ and $\frac{dy}{dt}$ and hence find $\frac{dy}{dx}$. What is the purpose of the parameters?

(c) By eliminating the parameter t from the x and y coordinates, obtain a cartesian equation for the curve. Comment on the type of the curve.

Solution 8

(a) When $t = 0$; $x = 0$, $y = 0$

$t = \frac{1}{4}$; $x = \frac{3}{16}$, $y = \frac{45}{64}$

$t = \frac{1}{2}$; $x = \frac{3}{4}$, $y = \frac{9}{8}$

$t = 1$; $x = 3$, $y = 0$

$t = 2$; $x = 12$, $y = -18$

$t = 3$; $x = 27$, $y = -72$

$t = -1$; $x = 3$ $y = 0$

$t = -2$; $x = 12$, $y = 18$

$t = -3$; $x = 27$, $y = 72$

$t = -\frac{1}{2}$; $x = \frac{3}{4}$, $y = \frac{-3}{2} + \frac{3}{8}$

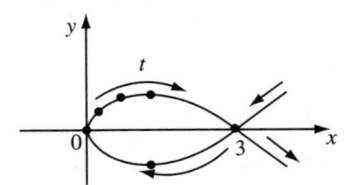

(b) $\frac{dx}{dt} = 6t$ $\frac{dy}{dt} = 3 - 9t^2$

$\frac{dy}{dx} = \frac{3 - 9t^2}{6t}$

$\frac{dy}{dx} = 0$ for turning points

$9t^2 = 3 \Rightarrow t = \pm \frac{1}{\sqrt{3}}$

This helps us find the turning points.

The use of the parameters, makes the sketching much easier to establish.

(c) $t^2 = \frac{x}{3} \Rightarrow t = \pm\sqrt{\frac{x}{3}}$

$\Rightarrow y = 3\left(\sqrt{\frac{x}{3}}\right) - 3\left(\sqrt{\frac{x}{3}}\right)^3$

$y = 3\frac{\sqrt{x}}{\sqrt{3}}\frac{\sqrt{3}}{\sqrt{3}} - 3\frac{x}{3}\sqrt{\frac{x}{3}}$

$y = \sqrt{3x} - x\sqrt{\frac{x}{3}}$

This is rather difficult to sketch or $y = -3\sqrt{\frac{x}{3}} + x\sqrt{\frac{x}{3}}$; when $t = -\sqrt{\frac{x}{3}}$.

Example 9

A curve has parametric equations
$x = 2\sin t$, $y = 3\cos t$, $0 \le t \le 2\pi$.

(a) Find an expression for $\frac{dy}{dx}$ in terms of the parameter t.

(b) Find the cartesian equation of the curve.

(c) Sketch the curve.

(d) Determine the equations of the tangents at $t = 0$ and $t = \pi$.

Solution 9

(a) $x = 2\sin t$ $y = 3\cos t$

$\frac{dx}{dt} = 2\cos t$ $\frac{dy}{dt} = -3\sin t$

$\frac{dy}{dx} = \frac{\frac{dy}{dt}}{\frac{dx}{dt}} = \frac{-3\sin t}{2\cos t} = -\frac{3}{2}\tan t$

(b) $\frac{x}{2} = \sin t \Rightarrow \frac{x^2}{4} = \sin^2 t$

$\frac{y}{3} = \cos t \Rightarrow \frac{y^2}{9} = \cos^2 t$

$\cos^2 t + \sin^2 t = 1 = \frac{x^2}{4} + \frac{y^2}{9}$

$\frac{x^2}{4} + \frac{y^2}{9} = 1$

(c) $\dfrac{x^2}{4} + \dfrac{y^2}{9} = 1$

when $x = 0$, $y = \pm 3$

when $y = 0$, $x = \pm 2$

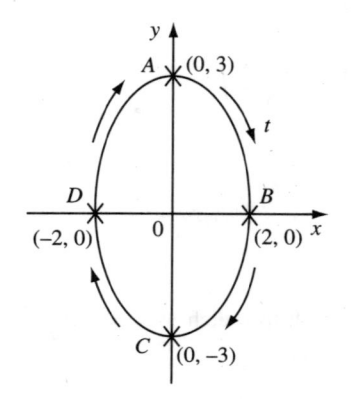

t increases from A to B to C to D and back to A.

(d) The gradient at $t = 0$

$\dfrac{dy}{dx} = -3 \sin 0 = 0$

$y = mx + c$

$y = c = 3$

$\therefore \boxed{y = 3}$

The gradient at $t = \pi$

$\dfrac{dy}{dx} = -3 \sin \pi = 0$

$y = c = 3 \cos \pi = -3$

$\boxed{y = -3}$

Example 10

Differentiate the function implicitly

$xy + x^2y + xy^2 = 5$ ... (1)

Solution 10

Differentiate (1) with respect to x

$1 \cdot y + x\dfrac{dy}{dx} + 2xy + x^2\dfrac{dy}{dx} + y^2 + 2xy\dfrac{dy}{dx} = 0$

$\dfrac{dy}{dx}(x + x^2 + 2xy) = -y - 2xy - y^2$

$\dfrac{dy}{dx} = \dfrac{-(y + 2xy + y^2)}{x + x^2 + 2xy}$

$= \dfrac{-y(1 + 2x + y)}{x(1 + x + 2y)}$ ... (2)

Differentiate (1) with respect to y

$\dfrac{dx}{dy} \cdot y + x \cdot 1 + 2x\dfrac{dx}{dy}y + x^2 \cdot 1 + \dfrac{dx}{dy} \cdot y^2 + 2xy = 0$

$\dfrac{dx}{dy}(y + 2xy + y^2) = -(x + x^2 + 2xy)$

$\dfrac{dx}{dy} = \dfrac{-x(1 + x + 2y)}{y + 2xy + y^2}$

$= \dfrac{-x(1 + x + 2y)}{y(1 + 2x + y)}$ ... (3)

Equations (2) and (3) are identical.

FP2, FP3

Intrinsic coordinates and radius of curvature.

Consider a curve with equation $y = f(x)$

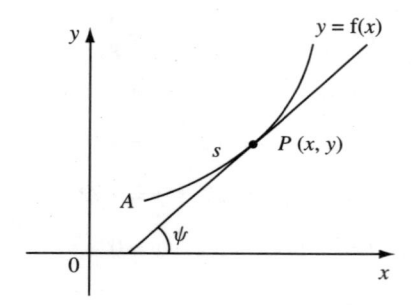

The tangent at $P(x, y)$ of the curve $y = f(x)$, makes an angle ψ, called the *gradient angle*, with the positive x-axis and $AP = s$, the length of the arc.

As P changes position on the curve so does s and ψ.

$\dfrac{dy}{dx} = \tan \psi$

the gradient at P

$(ds)^2 = (dx)^2 + (dy)^2$

$\left(\dfrac{ds}{dx}\right)^2 = \left(\dfrac{dx}{dx}\right)^2 + \left(\dfrac{dy}{dx}\right)^2$

$\left(\dfrac{ds}{dx}\right)^2 = 1 + \left(\dfrac{dy}{dx}\right)^2$

$\sec^2 \psi = 1 + \tan^2 \psi$

$\therefore \dfrac{ds}{dx} = \sec \psi$

$$\frac{dx}{ds} = \frac{1}{\frac{ds}{dx}} = \frac{1}{\sec\psi} = \cos\psi$$

$$\frac{dy}{ds} = \frac{1}{\frac{ds}{dy}} = \frac{1}{\operatorname{cosec}\psi} = \sin\psi$$

since

$$(ds)^2 = (dx)^2 + (dy)^2$$

$$\left(\frac{ds}{dy}\right)^2 = \left(\frac{dx}{dy}\right)^2 + \left(\frac{dy}{dy}\right)^2$$

$$\operatorname{cosec}^2\psi = \cot^2\psi + 1$$

$$\therefore \quad \frac{dy}{dx} = \tan\psi, \quad \frac{dx}{ds} = \cos\psi \text{ and } \frac{dy}{ds} = \sin\psi$$

$$y = f(x) \text{ cartesian form}$$

$$s = g(\psi) \text{ intrinsic form}$$

$$\frac{ds}{d\psi} = \text{radius of curvature} = \rho$$

$$= \text{the rate of change of } s \text{ with respect to } \psi$$

$$\rho = \frac{ds}{d\psi}$$

$$\frac{dy}{dx} = \tan\psi$$

$$\frac{d^2y}{dx^2} = \frac{d}{dx}(\tan\psi) = \sec^2\psi \frac{d\psi}{dx}$$

$$\therefore \quad \frac{d\psi}{dx} = \frac{\frac{d^2y}{dx^2}}{\sec^2\psi} \quad \text{or} \quad \frac{dx}{d\psi} = \frac{\sec^2\psi}{\frac{d^2y}{dx^2}}$$

$$\rho = \frac{ds}{d\psi} = \frac{\frac{ds}{dx}}{\frac{d\psi}{dx}}$$

$$= \frac{\frac{1}{\cos\psi}}{\frac{d^2y}{dx^2} \cdot \frac{1}{\sec^2\psi}} = \frac{\sec^3\psi}{\frac{d^2y}{dx^2}}$$

$$1 + \tan^2\psi = \sec^2\psi$$

$$\sec\psi = \sqrt{1 + \tan^2\psi}$$

$$\boxed{\rho = \frac{\left[1 + \left(\frac{dy}{dx}\right)^2\right]^{\frac{3}{2}}}{\frac{d^2y}{dx^2}}}$$

For parametric coordinates

$$\frac{dy}{dx} = \frac{\frac{dy}{dt}}{\frac{dx}{dt}} = \frac{\dot{y}}{\dot{x}}$$

$$\frac{d^2y}{dx^2} = \frac{d}{dx}\left(\frac{\dot{y}}{\dot{x}}\right) = \frac{\dot{x}\frac{d(\dot{y})}{dt}\frac{dt}{dx} - \dot{y}\frac{d(\dot{x})}{dt}\frac{dt}{dx}}{\dot{x}^2}$$

$$= \frac{\dot{x}\ddot{y}\frac{1}{\dot{x}} - \dot{y}\ddot{x}\frac{1}{\dot{x}}}{\dot{x}^2} = \frac{\dot{x}\ddot{y} - \dot{y}\ddot{x}}{(\dot{x})^3}$$

$$\rho = \frac{\left[1 + \left(\frac{\dot{y}}{\dot{x}}\right)\right]^{\frac{3}{2}}}{\frac{\dot{x}\ddot{y} - \dot{y}\ddot{x}}{\dot{x}^3}} = \frac{(\dot{x}^2 + \dot{y}^2)^{\frac{3}{2}}}{\dot{x}\ddot{y} - \dot{y}\ddot{x}}$$

$$\boxed{\rho = \frac{(\dot{x}^2 + \dot{y}^2)^{\frac{3}{2}}}{\dot{x}\ddot{y} - \dot{y}\ddot{x}}}$$

Example 11

The parametric equations of a curve are

$$x = ct \text{ and } y = \frac{c}{t}$$

Find the radius of curvature at any point on the curve in terms of t, hence find ρ when $c = 2$ and $t = 1$.

Solution 11

$$\rho = \frac{(\dot{x}^2 + \dot{y}^2)^{\frac{3}{2}}}{\dot{x}\ddot{y} - \dot{y}\ddot{x}}$$

$$x = ct \qquad y = \frac{c}{t} = ct^{-1}$$

$$\dot{x} = c \qquad \dot{y} = -ct^{-2} = -\frac{c}{t^2}$$

$$\ddot{x} = 0 \qquad \ddot{y} = 2ct^{-3} = \frac{2c}{t^3}$$

substituting in (1)

$$\rho = \frac{\left(c^2 + \left(\frac{-c}{t^2}\right)^2\right)^{\frac{3}{2}}}{c\frac{2c}{t^3} - \left(\frac{-c}{t^3}\right)0}$$

$$\rho = \frac{\left(\frac{c^2t^4 + c^2}{t^4}\right)^{\frac{3}{2}}}{\frac{2c^2}{t^3}} = \frac{\frac{c^3(t^4+1)^{\frac{3}{2}}}{t^6}}{\frac{2c^2}{t^3}}$$

$$\rho = \frac{c(t^4 + 1)^{\frac{3}{2}}}{2t^3}$$

$$\rho = \frac{2 \times 2\sqrt{2}}{2 \times 1} = 2\sqrt{2}$$

Example 12

The parametric equation of a curve are $x = \cos t + t \sin t$, $y = \sin t - t \cos t$, $0 < t < \frac{\pi}{2}$

Find the radius of curvature at any point on the curve in terms of t, hence find ρ when $t = 2$.

Solution 12

$x = \cos t + t \sin t$

$\dot{x} = -\sin t + \sin t + t \cos t$

$\dot{x} = t \cos t$ ... (1)

$\ddot{x} = \cos t - t \sin t$ ... (2)

$y = \sin t - t \cos t$

$\dot{y} = \cos t - \cos t - t(-\sin t)$

$= t \sin t$ ... (3)

$\ddot{y} = \sin t + t \cos t$ ... (4)

$\rho = \dfrac{(\dot{x}^2 + \dot{y}^2)^{\frac{3}{2}}}{\dot{x}\ddot{y} - \dot{y}\ddot{x}}$

substituting (1), (2), (3) and (4) into ρ

$\rho = \dfrac{(t^2 \cos^2 t + t^2 \sin^2 t)^{\frac{3}{2}}}{t \cos t(\sin t + t \cos t) - t \sin t(\cos t - t \sin t)}$

$= \dfrac{t^3}{t \sin t \cos t + t^2 \cos^2 t - t \sin t \cos t + t^2 \sin^2 t}$

$= \dfrac{t^3}{t^2} = t$

$\boxed{\rho = 2}$ the radius of curvature.

Example 13

A cycloid has parametric equations $x = t - \sin t$, $y = 1 - \cos t$.

(a) Determine the intrinsic equation of this curve.

(b) Determine the radius of curvature at any point on the curve, hence find its value at $t = \pi$.

Solution 13

(a) $\dot{x} = 1 - \cos t$ $\qquad \ddot{x} = \sin t$

$\dot{y} = \sin t$ $\qquad \ddot{y} = \cos t$

$\dfrac{dy}{dx} = \dfrac{\dot{y}}{\dot{x}} = \dfrac{\sin t}{1 - \cos t}$

$= \dfrac{2 \sin \frac{t}{2} \cos \frac{t}{2}}{1 - 2\cos^2 \frac{t}{2} + 1}$

$= \dfrac{2 \sin \frac{t}{2} \cos \frac{t}{2}}{2(1 - \cos^2 \frac{t}{2})} = \dfrac{2 \sin \frac{t}{2} \cos \frac{t}{2}}{2 \sin^2 \frac{t}{2}}$

$= \cot \dfrac{t}{2} = \tan \psi$

$\psi = \dfrac{\pi}{2} - \dfrac{t}{2}$

$s = \text{length of arc} = \displaystyle\int_0^t \sqrt{\dot{x}^2 + \dot{y}^2}\, dt$

$= \displaystyle\int_0^t \sqrt{(1 - \cos t)^2 + \sin^2 t}\, dt$

$= \displaystyle\int_0^t \sqrt{1 - 2\cos t + \cos^2 t + \sin^2 t}\, dt$

$= \displaystyle\int_0^t \sqrt{2 - 2\cos t}\, dt$

$= \displaystyle\int_0^t \sqrt{2 - 2\left(1 - 2\sin^2 \dfrac{t}{2}\right)}\, dt$

$= \displaystyle\int_0^t 2 \sin \dfrac{t}{2}\, dt = \left[-4 \cos \dfrac{t}{2}\right]_0^t$

$= -4 \cos \dfrac{t}{2} + 4 \cos 0$

$s = 4\left(1 - \cos \dfrac{t}{2}\right) = 4\left[1 - \cos\left(\dfrac{\pi}{2} - \psi\right)\right]$

$= 4(1 - \sin \psi)$

The intrinsic equation of the cycloid is
$s = 4(1 - \sin \psi)$

(b) $\rho = \dfrac{(\dot{x}^2 + \dot{y}^2)^{\frac{3}{2}}}{\dot{x}\ddot{y} - \dot{y}\ddot{x}}$

$\rho = \dfrac{\left[(1 - \cos t)^2 + \sin^2 t\right]^{\frac{3}{2}}}{(1 - \cos t)\cos t - \sin t(\sin t)}$

$= \dfrac{(1 - 2\cos t + \cos^2 t + \sin^2 t)^{\frac{3}{2}}}{\cos t - \cos^2 t - \sin^2 t}$

$= \dfrac{(2 - 2\cos t)^{\frac{3}{2}}}{\cos t - 1}$

$= -\dfrac{2^{\frac{3}{2}}(1 - \cos t)^{\frac{3}{2}}}{1 - \cos t}$

$= -2^{\frac{3}{2}}(1 - \cos t)^{\frac{1}{2}}$

$= -2\sqrt{2}\left(1 - 1 + 2\sin^2 \dfrac{t}{2}\right)^{\frac{1}{2}}$

$= -2\sqrt{2}\left(2\sin^2 \dfrac{t}{2}\right)^{\frac{1}{2}}$

$= -2\sqrt{2}\sqrt{2}\sin \dfrac{t}{2}$

$= -4\sin \dfrac{t}{2}$

at $t = \pi$ $\boxed{\rho = -4}$

FP2

Inverse Trigonometric Functions are shown adequately in pages 17 and 18.

Inverse Hyperpolic Functions are shown adequately in pages 29 to 32.

FP3

Numerical Methods for the solution of Differential Equations Polynomial Approximations Using Taylor Series are shown in pages 63 to 67.

4. Answers

Exercises 1

1. anx^{n-1}

2. (i) 0
   (ii) 1
   (iii) $-2x$
   (iv) $6x^2$
   (v) $5 + \dfrac{2}{x^2} - \dfrac{2}{x^3}$

3. (i) 3
   (ii) 0
   (iii) $-\dfrac{3}{x^2}$
   (iv) $-2x - 3x^2 - 4x^3$
   (v) $-\dfrac{3}{2x^{\frac{3}{2}}}$
   (vi) $-\dfrac{1}{x^2} - \dfrac{8}{x^3} + \dfrac{9}{x^4}$
   (viii) $6t - 5$
   (viii) $6t - 5$
   (ix) $-\dfrac{1}{t^2} + 1 - 2t$
   (x) $10y - 15y^2 - 35y^4$

4. (i) $4x^3 - 4x$
   (ii) $4x^3 + 3x^2 - 9$
   (iii) $24t^7 - 25t^4$
   (iv) $81x^{26}$
   (v) $(x^2 - 1)(x^3 - 2) + 2x^2(x^3 - 2) + 3x^3(x^2 - 1)$

5. (i) $-\dfrac{x(3x^5 + 2)}{(x^5 - 1)^2}$
   (ii) $\dfrac{6t^2}{(t^3 + 1)^2}$
   (iii) $-\dfrac{3(1 + 3r^4)}{(r^4 - 1)^2}$
   (iv) $\dfrac{3}{5}x^2 - \dfrac{6}{5}x + \dfrac{2}{5}$
   (v) $\dfrac{3}{(x + 2)^2}$

6. (i) $\dfrac{3}{2(3x + 1)^{\frac{1}{2}}}$
   (ii) $-\dfrac{1}{2(x - 1)^{\frac{3}{2}}}$
   (iii) $9x^2 (x^3 - 1)^2$
   (iv) $6x(x^2 - 1)^2 (x^3 + 1)^3 (3x^3 - 2x + 1)$
   (v) $\dfrac{10}{3}x(5x^2 - 7)^{-\frac{2}{3}}$

8. (a) (i) $-\dfrac{c^2}{x^2}$
   (ii) $\dfrac{b^2}{a^2}\dfrac{x}{y}$
   (iii) $\dfrac{x}{y}\dfrac{a^2}{b^2}$
   (iv) $-\dfrac{x}{y}$
   (v) $-\dfrac{x}{y}\dfrac{a^2}{b^2}$

(vi) $-\dfrac{y}{x+2y}$

(vii) $\dfrac{3y-2x}{2y-3x+5}$

(viii) $-\dfrac{x+g}{y+f}$

(ix) $\dfrac{2a}{y}$

(x) $-\dfrac{2x}{5}$

(b) (i) $-\dfrac{x}{y}$

(ii) $\dfrac{ya^2}{xb^2}$

(iii) $\dfrac{ya^2}{xb^2}$

(iv) $-\dfrac{y}{x}$

(v) $-\dfrac{ya^2}{xb^2}$

(vi) $-\dfrac{x+2y}{y}$

(vii) $\dfrac{3x-2y-5}{2x-3y}$

(viii) $-\dfrac{f+y}{x+g}$

(ix) $\dfrac{y}{2a}$

(x) $-\dfrac{5}{2x}$

9. (i) -11

(ii) -1

(iii) -1

(iv) $-\dfrac{1}{5}$

(v) $\dfrac{2}{3}$

(vi) $\pm\dfrac{1}{4}$

(vii) 1

(viii) -6

(ix) $\pm\dfrac{1}{\sqrt{3}}$

(x) $-\dfrac{1}{2}$

10. (i) $an(n-1)x^{n-2}$

(ii) $-\dfrac{1}{2}x^{-\frac{3}{2}}+\dfrac{1}{2}x^{-\frac{1}{2}}+\dfrac{2}{3}x^{-\frac{1}{3}}$

(iii) $\dfrac{2x^3}{(x^4-1)^{\frac{1}{2}}}$

(iv) 1

(v) $\dfrac{9x^2}{(x^3+2)^2}$

(vi) $-\dfrac{3}{5(1-3x)^{\frac{4}{5}}}$

(vii) $2x(x-1)^{\frac{1}{2}}+\dfrac{x^2}{2}(x-1)^{-\frac{1}{2}}$

(viii) $\dfrac{2x-y}{x-2y}$

(ix) $\dfrac{1-6x}{6x+1}$

(x) $-\dfrac{3x}{5y}$

11. (i) $8(2x+5)^3$

(ii) $-4(x-1)^{-5}$

(iii) $\dfrac{3}{2}(1+3x)^{-\frac{1}{2}}$

(iv) $5(2+6x)(1+2x+3x^2)^4$

(v) $\dfrac{1}{2}(5)(5x-7)^{\frac{1}{2}}$

(vi) $-\dfrac{3}{2(3x-2)^{\frac{3}{2}}}$

(vii) $\dfrac{7}{2(1+7x)^{\frac{1}{2}}}$

(viii) $-\dfrac{4x}{(1+4x^2)^{\frac{3}{2}}}$

(ix) $2(10x-7)(5x^2+7x-3)$

(x) $\dfrac{x^4+4x^2+2x+7}{(x^2-1)^2}$

12. (i) $-\sqrt{\dfrac{y}{x}}$

(ii) $-\dfrac{4x+3y}{3x}$

(iii) $-\dfrac{x}{y}$

(iv) $\dfrac{2}{5y}$

(v) $-\dfrac{2}{3}$

(vi) $-\dfrac{2x}{3y}$

13. (i) $\sqrt{3}-1$

(ii) $-\dfrac{7}{3}$

(iii) ± 0.3535

(iv) $\pm\dfrac{\sqrt{5}}{5} = \pm 0.447$

(v) $-\dfrac{2}{3}$

(vi) ± 0.241

14. (i) $-\dfrac{x^2+14x-2}{(x^2-x+5)^2}$

(ii) $\dfrac{1}{(1-x)^2}$

(iii) $-\dfrac{1}{2x^{\frac{3}{2}}}$

(iv) $\dfrac{-(4x+3)}{(2x^2+3x+4)^2}$

(v) $\dfrac{1}{2\sqrt{x}} + \dfrac{1}{2\sqrt{x^3}}$

15. (i) $20x^3+5$

(ii) $\dfrac{2}{(y+1)^2}$

(iii) $-\dfrac{1}{z^2} - \dfrac{2}{z^3} - \dfrac{3}{z^4}$

(iv) $\dfrac{3}{2}t^2 - \dfrac{1}{2t^2}$

(v) $5u^4+3u^2+4u$

16. (i) $-\dfrac{1}{2x^{\frac{3}{2}}} + \dfrac{1}{x^2} - \dfrac{2}{x^3}$

(ii) $3x^2+12x+11$

(iii) $\dfrac{1}{\sqrt{x}(\sqrt{x}+1)^2}$

(iv) $\dfrac{10}{x^3}\left(1-\dfrac{1}{x^2}\right)^4$

(v) $-\dfrac{y}{x}$

17. (i) $-\dfrac{4}{25}$

(ii) -25.

Exercises 2

1. (i) $3\cos x$

(ii) $2\sin x$

(iii) $2\sec^2 2x$

2. (i) $\cos x$

(ii) $-2\sin x$

(iii) $3\sec^2 x$

(iv) $-4\csc^2 x$

(v) $-5\cot x \csc x$

(vi) $6\tan x \sec x$

3. (i) $\cos 2x$

   (ii) $-\sin \dfrac{x}{3}$

   (iii) $\sec^2 \dfrac{x}{4}$

   (iv) $-\cot 5x \operatorname{cosec} x \, 5x$

   (v) $49 \sec 7x \tan 7x$

   (vi) $-6 \operatorname{cosec}^2 3x$

4. (a) (i) 1

   (ii) 0

   (iii) 1

   (iv) $-\infty$

   (v) 0

   (vi) $-\infty$

   (b) (i) 0

   (ii) -0.259

   (iii) 1.04

   (iv) 1.414

   (v) -69.3

   (vi) -12

   (c) (i) 0

   (ii) -0.707

   (iii) 1.45

   (iv) -1.414

   (v) -69.3

   (vi) -12

5. (i) $\sin x + x \cos x$

   (ii) $2x \sin^2 x + 2x^2 \sin x \cos x$

   (iii) $2 \tan x \sec^2 x$

   (iv) $6 \sec^2 x \tan^2 x + 3 \sec^2 x \sec^2 x$

   (v) $-15 \operatorname{cosec}^3 x \cot x$

   (vi) $-4 \cot^3 x \operatorname{cosec}^2 x$

6. (i) $\dfrac{3 \cos t}{2 \sin^{\frac{1}{4}} t}$

   (ii) $-\dfrac{15}{2} \tan^{\frac{3}{2}} 3t \sec^2 3t$

   (iii) $-\dfrac{1}{2} \sqrt{\operatorname{cosec} xt} \cot t$

7. (i) $2x\sqrt{\cos x} - \dfrac{1}{2}x^2 \dfrac{\sin x}{\sqrt{\cos x}}$

   (ii) $\sqrt{\sin x} + \dfrac{1}{2}x \dfrac{\cos x}{\sqrt{\sin x}}$

   (iii) $\sqrt{\tan x} + \dfrac{1}{2}x \dfrac{\sec^2 x}{\sqrt{\tan x}}$

8. $\dfrac{y}{1 + x^2 y^2 - x}$

9. $-\dfrac{\dfrac{x \cos^{-1} x}{\sqrt{1-x^2}} + 1}{\left(\cos^{-1} x\right)^2}$

10. $\dfrac{1}{1-x^2} + \dfrac{x \sin^{-1} x}{\left(1-x^2\right)^{\frac{3}{2}}}$

11. $\dfrac{\cot^{-1} x + \tan^{-1} x}{(1+x^2)\left(\cot^{-1} x\right)^2}$

12. $-\dfrac{\sec^{-1} x + \operatorname{cosec}^{-1} x}{x \left(x^2 - 1\right)^{\frac{1}{2}} \left(\sec^{-1} x\right)^2}$

13. $-\dfrac{\sin^{-1} x + \cos^{-1} x}{\left(1-x^2\right)^{\frac{1}{2}} \left(\sin^{-1} x^2\right)}$

14. (i) $\dfrac{9}{\sqrt{1-9x^2}}$

    (ii) $-\dfrac{2}{1+4x^2}$

    (iii) $-\dfrac{20}{\sqrt{1-16x^2}}$

15. (i) $\cos x \cos^{-1} x - \sin x \dfrac{1}{(1-x^2)^{\frac{1}{2}}}$

    (ii) $-2 \sin x \sin^{-1} x + 2 \cos x \dfrac{1}{\sqrt{1-x^2}}$

    (iii) $3 \sec^2 x \cot^{-1} x - \dfrac{3 \tan x}{1+x^2}$.

Exercises 3

1. (i) $1 + \dfrac{x}{1!} + \dfrac{x^2}{2!} + \cdots$

   (ii) $1 - \dfrac{x}{1!} + \dfrac{x^2}{2!} - \cdots$

   (iii) $1 + \dfrac{2x}{1!} + \dfrac{4x^2}{2!} + \cdots$

   (iv) $1 - \dfrac{3x}{1!} + \dfrac{9x^2}{2!} - \cdots$

   (i) e^x

   (ii) $-e^{-x}$

   (iii) $2e^{2x}$

   (iv) $-3e^{-3x}$

2. (i) $3e^x$

   (ii) $-3e^{-3x}$

   (iii) $2xe^{x^2}$

   (iv) $-6xe^{-3x^2}$

   (v) ane^{ax}

   (vi) $\dfrac{1}{2}e^{\frac{x}{2}}$

   (vii) $-\dfrac{1}{4}e^{-\frac{x}{2}}$

3. (i) $6xe^{x^2} + 6x^3 e^{x^2}$

   (ii) $e^x \sin x + e^x \cos x$

   (iii) $-3e^{-3x} \cos 3x - 3e^{-3x} \sin 3x$

   (iv) $-3e^{-x} \sec x + 3e^{-x} \sec x \tan x$

   (v) $3e^{3x}(x^3 + 3) + e^{3x}(3x^2)$

4. (i) $\dfrac{e^x \sin x - e^x \cos x}{\sin^2 x}$

   (ii) $\dfrac{\sec^2 x e^{2x} - 2 \tan x e^{2x}}{e^{4x}}$

   (iii) $\dfrac{3e^{3x} \cos x + 3^{3x} \cos x \cot x + e^{3x} \operatorname{cosec} x}{\cot^2 x}$

5. (i) $3x^2 e^{3x}$

   (ii) $-2e^{-2x}(x^2 - 1) + 2xe^{-2x}$

   (iii) $6e^{2x} \cos 2x - 6e^{2x} \sin 2x$

   (iv) $\dfrac{e^x (\tan x - \sec^2 x)}{\tan^2 x}$

   (v) $3e^{3x} \cos(e^{3x})$

   (vi) $6xe^{x^2}$

   (vii) $2e^{x^2} \sin x^2 + e^{x^2} 2x \cos x^2$

   (viii) $45 e^{15x}$

   (ix) $e^x + e^{-x}$

   (x) $2e^{2x} - 2e^{-2x}$

   (xi) $-\dfrac{4}{(e^x - e^{-x})^2}$

   (xii) $-e^{-x} e^{e^{-x}}$

   (xiii) $\dfrac{-6(e^{2x} - e^{-2x})}{(e^{2x} + e^{-2x})^4}$

   (xiv) $3(\log_e a) a^x$

   (xv) $2(\log_e 3) 3^x$

6. (i) $ne^{nx} \sin kx + e^{nx} k \cos kx$

   (ii) $-me^{-mx} \cos nx - ne^{-nx} \sin nx$

   (iii) $\dfrac{ae^{ax} \cos bx + be^{ax} \sin bx}{\cos^2 bx}$

   (iv) $2e^{2x} \cos 5x - 5e^{2x} \sin 5x$

   (v) $e^x \sin 5x + 5e^x \cos 5x$

7. (i) $\dfrac{1}{t} e^{-\frac{1}{t}}$

   (ii) $\dfrac{1}{2\sqrt{u}} e^{u^{\frac{1}{2}}}$

   (iii) $-\cos x \, e^{-\sin x}$

8. (i) $-e^{-x}$

   (ii) $2e^{2x}$

10. $-\dfrac{1}{x^2}e^{\frac{1}{x}}$

11. Graph

12. Graph

13. $\dfrac{d^2y}{dx^2} = 13\,e^{2x}\cos(3x+\alpha)$ where $\alpha = \tan^{-1}\dfrac{5}{12}$

14. Proof

15. Proof

16. (i) $\dfrac{dy}{dx} = \dfrac{e^{x^2}(2x\cos x + \sin x)}{\cos^2 x}$

   (ii) $\dfrac{dy}{dx} = e^{-x^2}(\cos x - 2x\sin x)$

   (iii) $\dfrac{dy}{dx} = 3(e^x \tan 2x)^2 e^x(\tan 2x + 2\sec^2 2x)$

17. (i) $\dfrac{dy}{dx} = e^{-\frac{1}{x}}\left(\dfrac{1}{x^2}\sin x + \cos x\right)$

   (ii) $\dfrac{dy}{dx} = 2e^{-\frac{1}{x^2}}\left(\dfrac{1}{x^3}\cos 2x - \sin 2x\right)$

18. $\dfrac{dy}{dx} = -\dfrac{2}{3}\dfrac{e^{2x}}{\sin 3y}$

19. $\dfrac{dy}{dx} = \dfrac{2\sec^2 x}{e^{\sqrt{y}}}\sqrt{y}$

20. $\dfrac{dy}{dx} = -\dfrac{2}{3}e^{-2x}(\cos x - 2x\sin x)$

21. $\dfrac{d^2y}{dx^2} = 2\cot x - 4x\,\text{cosec}^2 x + 2x^2\,\text{cosec}^2 x \cot x$

22. $\dfrac{d^2y}{dx^2} = 2\cot x - 4x\,\text{cosec}^2 x + 2x^2\,\text{cosec}^2 x \cot x$

Exercises 4

1. (i) $\dfrac{dy}{dx} = \dfrac{1}{x\log_e 10}$

   (ii) $\dfrac{dy}{dx} = \dfrac{3}{x}$

   (iii) $\dfrac{dy}{dx} = (3x)^x(\ln 3x + 1)$

   (iv) $\dfrac{dy}{dx} = 7^x \ln 7$

   (v) $\dfrac{dy}{dx} = \dfrac{1}{x}$

2. (i) $\dfrac{dy}{dx} = (\cos x)^x\left(\ln \cos x - x\dfrac{\sin x}{\cos x}\right)$

   (ii) $\dfrac{dy}{dx} = (\cot x)^x\left(\ln \cot x - x\dfrac{\text{cosec}^2 x}{\cot x}\right)$

   (iii) $\dfrac{dy}{dx} = (x+1)^x\left[\ln(x+1) + \dfrac{x}{1+x}\right]$

   (iv) $\dfrac{dy}{dx} = (x^2+1)^{\frac{1}{2}}(x^3-1)^{-\frac{1}{2}}(x^4+1)^{-\frac{1}{2}}$
   $\times \left[\dfrac{x}{x^2+1} - \dfrac{3}{2}\dfrac{x^2}{x^3-1} - \dfrac{2x^3}{x^4+1}\right]$

   (v) $\dfrac{dy}{dx} = (x-1)^{\frac{1}{3}}(x+1)^{-\frac{1}{3}}(x+2)^{-\frac{1}{3}}$
   $\times \left[\dfrac{1}{3(x-1)} - \dfrac{1}{3(x+1)} - \dfrac{1}{3(x+2)}\right]$

3. (i) $\dfrac{dy}{dx} = \dfrac{1}{5x}$

   (ii) $\dfrac{dy}{dx} = \dfrac{1}{1-x} - \dfrac{1}{x}$

   (iii) $\dfrac{dy}{dx} = 2x\ln x + x$

4. (i) $\dfrac{dy}{dx} = -\dfrac{3}{x^4}\ln 3x + \dfrac{1}{x^4}$

   (ii) $\dfrac{dy}{dx} = 1 - \dfrac{1}{x}$

   (iii) $\dfrac{dy}{dx} = \dfrac{\ln x - 1}{(\ln x)^2}$

   (iv) $\dfrac{dy}{dx} = \dfrac{\frac{1}{x}\sin 2x - \ln 2x(2\cos 2x)}{\sin^2 2x}$

   (v) $\dfrac{dy}{dx} = \dfrac{\sec^2 x \ln\left|\frac{1}{x}\right| + \frac{1}{x}\tan x}{\left[\ln\left|\frac{1}{x}\right|\right]^2}$

5. $\dfrac{dy}{dx} = e^x \ln x + \dfrac{e^x}{x}$

6. $\dfrac{dy}{dx} = \cos x\,e^{\sin x}\cos(\ln x) + e^{\sin x}(-\sin(\ln x))\dfrac{1}{x}$

7. $\dfrac{dy}{dx} = -\sin x\,e^{\cos x}\ln(\sin x) + e^{\cos x}\dfrac{1}{\sin x}\cos x$

8. $\dfrac{dy}{dx} = 10\tan(5x-1)$

9. (i) $\dfrac{dy}{dx} = \sqrt{x(x-1)(x+2)}$

    $\times \left[\dfrac{1}{2}\left(\dfrac{1}{x} + \dfrac{1}{x-1} + \dfrac{1}{x+2} \right) \right]$

    (ii) $\dfrac{dy}{dx} = \sqrt{(x+1)(x+3)(x+4)}$

    $\times \left[\dfrac{1}{2}\left(\dfrac{1}{x+1} + \dfrac{1}{x+3} + \dfrac{1}{x+4} \right) \right]$

10. $\dfrac{dy}{dx} = \sqrt{\dfrac{3x^3 - 4}{5x^3 + 7}} \left[\dfrac{1}{2}\left(\dfrac{9x^2}{3x^3 - 4} - \dfrac{15x^2}{5x^3 + 7} \right) \right]$

Exercises 5

1. (i) $\dfrac{dy}{dx} = \cosh x$

    (ii) $\dfrac{dy}{dx} = \sinh x$

    (iii) $\dfrac{dy}{dx} = \dfrac{1}{2}\cosh\dfrac{1}{2}x$

2. Proof

3. (i) $\dfrac{dy}{dx} = 2\sec^2 2x \coth 3x$

    $-3\tan 2x \operatorname{cosech}^2 3x$

    (ii) $\dfrac{dy}{dx} = 3\cosh 3x \cot 2x$

    $-2\sinh 2x \csc^2 2x$

    (iii) $\dfrac{dy}{dx} = \dfrac{1}{x^2}\coth\dfrac{1}{x}\operatorname{cosech}\dfrac{1}{x}$

    (iv) $\dfrac{dy}{dx} = -2x\tanh x^2 \operatorname{sech} x^2$

    (v) $\dfrac{dy}{dx} = \dfrac{15}{2}\sinh^4\dfrac{x}{2}\cosh\dfrac{x}{2}$

    (vi) $\dfrac{dy}{dx} = -\dfrac{1}{2}\dfrac{\operatorname{cosech}^2 x}{\coth^{\frac{1}{2}} x}\sinh^{\frac{3}{2}} + \dfrac{3}{2}\cosh^{\frac{3}{2}} x$

4. (i) $\dfrac{dy}{dx} = \dfrac{3}{x(1+x^2)^{\frac{1}{2}}}$

    (ii) $\dfrac{dy}{dx} = 2x \dfrac{1}{(x^4 - 1)^{\frac{1}{2}}}$

    (iii) $\dfrac{dy}{dx} = 5(2x-3) \dfrac{1}{[(x^2 - 3x + 2)^2 - 1]^{\frac{1}{2}}}$

(iv) $\dfrac{dy}{dx} = -\coth\dfrac{x}{2}\operatorname{cosech}^2\dfrac{x}{2}\operatorname{sech}^2\dfrac{x}{3}$

$-\dfrac{2}{3}\operatorname{cosech}^2\dfrac{x}{2}\operatorname{sech}^2\dfrac{x}{3}\tanh\dfrac{x}{3}$

5. $\dfrac{d^2 y}{dx^2} = \dfrac{1}{x^2(1-4x^2)^{\frac{1}{2}}} - \dfrac{4}{(1-4x^2)^{\frac{3}{2}}}$

6. $\dfrac{d^2 y}{dx^2} = -\dfrac{x}{8\left(1 + \dfrac{1}{4}x^2\right)^{\frac{3}{2}}}$

7. $\dfrac{d(\operatorname{cosech}^{-1} 2x)}{dx} = -\dfrac{1}{x\sqrt{1+4x^2}}$

8. $\dfrac{d}{dx}(\tanh^{-1} 3x) = \dfrac{3}{1-9x^2}$

9. (i) 74.21

    (ii) 74.2

    (iii) 1.82×10^{-4}

10. (i) $\dfrac{dy}{dx} = 0.447, \dfrac{dy}{dx} = 0.707$

    (ii) $\dfrac{dy}{dx} = 0.577, \dfrac{dy}{dx} = \infty$

    (iii) $\dfrac{dy}{dx} = -0.333, \dfrac{dy}{dx} = \infty$

11. (i) $\dfrac{dy}{dx} = 2\cosh 2x \operatorname{cosech} 3x$

    $-3\sinh 2x \coth 3x \operatorname{cosech} 3x$

    (ii) $\dfrac{dy}{dx} = 3\cosh 3x$

    (iii) $\dfrac{dy}{dx} = e^x \cosh 2x + 2e^x \sinh 2x$

    (iv) $\dfrac{dy}{dx} = 5\coth 5x$

    (v) $\dfrac{dy}{dx} = -2\coth x \operatorname{cosech}^2 x e^{\coth^2 x}$

    (vi) $\dfrac{dy}{dx} = 3x^2 \coth^2 5x$

    $-15x^3 \coth^2 5x \operatorname{cosech}^2 5x$

    (vii) $\dfrac{dy}{dx} = -\dfrac{3}{2}\dfrac{\operatorname{cosech}^2 3x}{(\coth 3x)^{\frac{1}{2}}}$

(viii) $\dfrac{dy}{dx} = \sinh^3 x$

(ix) $\dfrac{dy}{dx} = 2\,\text{sech}^2 x (1 - 3\tanh^2 x)$

(x) $\dfrac{dy}{dx} = \sqrt{\dfrac{\cosh 2x + 1}{\cosh 2x - 1}} \left(-\dfrac{2}{\sinh 2x} \right)$

12. (i) $\dfrac{dy}{dx} = -3\,\text{cosech}\,3x$

(ii) $\dfrac{dy}{dx} = \dfrac{2\,\text{cosech}^2 2x}{\coth 2x (1 + \coth^2 2x)^{\frac{1}{2}}}$

(iii) $\dfrac{dy}{dx} = -\text{cosech}\,x$

(iv) $\dfrac{dy}{dx} = \dfrac{\cosh x}{1 - \sinh^2 x}$

(v) $\dfrac{dy}{dx} = \dfrac{2}{x(2 - 3x^2)}$

Exercises 6

1. $\dfrac{dy}{dx} = \dfrac{3}{2}\tanh t$

2. (a) $\dfrac{dy}{dx} = \cot\dfrac{t}{2},\ \dfrac{d^2 y}{dx^2} = -\dfrac{1}{4}\text{cosec}^4 \dfrac{t}{2}$

   (b) Sketch

3. (i) $\dfrac{dy}{dx} = -\dfrac{1}{t^2}$

   (ii) $\dfrac{d^2 y}{dx^2} = \dfrac{2}{ct^3}$. Sketch

4. Sketch

5. (i) $\dfrac{1}{t}$

   (ii) $-\dfrac{1}{2t^3}$. Sketch

6. Sketch

7. (i) $-\dfrac{3}{2}\sin 2t$

   (ii) $-3\cos 2t \dfrac{1}{2\cos t}$. Sketch

8. $\dfrac{dy}{dx} = \dfrac{\cos t + (t+3)\sin t}{-\sin t + (t+3)\cos t}$

9. $\dfrac{dy}{dx} = -2\cot\theta$

10. $\dfrac{dx}{dt} = 1 + e^t,\ \dfrac{dy}{dt} = 2 - 2e^{2t},\ \dfrac{dy}{dx} = \dfrac{2(1 - e^{2t})}{1 + e^t}$

    $\dfrac{d^2 y}{dx^2} = \dfrac{(-2e^{3t} - 4e^{2t} - 2e^t)}{(1 + e^t)^3}$

11. (i) $\tan t$

    (ii) $\dfrac{1}{3}\sec^2 t$

12. $\dfrac{dy}{dx} = \cot\dfrac{t}{2}$

13. (i) $\dfrac{4}{3}\cos t$

    (ii) $\dfrac{4}{9}$

14. $6\cos 2t - 6\cos 4t,\ -6\sin 2t - 6\sin 4t,$

    $\dfrac{dy}{dx} = -\dfrac{\sin 2t + \sin 4t}{\cos 2t - \cos 4t}$

15. $\dfrac{dy}{dx} = \tan\dfrac{3\theta}{2},\ \dfrac{d^2 y}{dx^2} = \dfrac{3}{8}\sec^3 \dfrac{3\theta}{2}\,\text{cosec}\,\dfrac{\theta}{2}$

16. $\dfrac{dy}{dx} = \dfrac{\pi}{4}$

17. $\dfrac{dy}{dx} = \cot\dfrac{\theta}{2},\ \dfrac{d^2 y}{dx^2} = -\dfrac{1}{12}\text{cosec}^4 \dfrac{\theta}{2}$

18. $\dfrac{dy}{dx} = -\text{cosech}\,t,$

    $\dfrac{d^2 y}{dx^2} = \text{sech}\,t(1 + \text{cosech}^2 t)^2,$

    $-\text{cosech}^3 t - \text{sech}\,t - 2\,\text{sech}\,t\,\text{cosech}^2 t$

    $-\text{sech}\,t\,\text{cosech}^4 t + \dfrac{\text{sech}\,t}{t - \tanh t}$

Exercises 7

1. (i) $6x - 5,\ 6$

   (ii) $1 - 12t + 21t^2,\ -12 + 42t$

   (iii) $6v + 5,\ 6$

   (iv) $6z - 1,\ 6$

2. (i) $\dfrac{3x^2 + 6x + 1}{(x+1)^2}$, $\dfrac{4}{(x+1)^3}$

   (ii) $e^x + \cos x$, $e^x - \sin x$

   (iii) $\dfrac{2e^{-x}\sin 2x - e^{-x}\cos 2x}{\cos^2 2x}$,

   $= \dfrac{\{5e^{-x}\cos^2 2x + 8e^{-x}\sin^2 2x - 4e^{-x}\sin 2x \cos 2x\}}{\cos^3 2x}$

   (iv) $6\cos 2x + 10\sin 2x$, $-12\sin 2x + 20\cos 2x$

   (v) $\sin 2x$, $2\cos 2x$

3. (i) 4 m/s

   (ii) 4 m/s when $t = 0$

   (iii) $v = 124$ m/s

   (iv) 18 m/s²

   (v) $x = 44$ m, $x = 200$ m

4. (i) 177 m/s

   (ii) 60 m/s²

   (iii) $s = 4.925$ m

   (iv) 57 m/s, $t = 1$

   (v) $v = 597$ m/s

5. (i) $40\,t$

   (ii) 40 m/s

   (iii) 8.66 s

   (iv) 40 m/s²

6. (i) $6\cos 2x + 10\sin 2x$, $-12\sin 2x + 20\cos 2x$

   (ii) $9x^2 - 4x + 1$, $18x - 4$

   (iii) $-8e^{-2x} - 15e^{3x}$, $16e^{-2x} - 45e^{3x}$

   (iv) $3e^{3x} + 3\sin 3x + 3\cos 3x$,
        $9e^{3x} + 9\cos 3x - 9\sin 3x$

   (v) $\dfrac{5}{x} + \sin 2x + 2x\cos 2x$,
       $-\dfrac{5}{x^2} + 2\cos 2x + 2\cos 2x - 4x\sin 2x$.

7. (i) $25t^4 - 16t^3 + 9t^2 - 4t + 1$,
       $100t^3 - 48t^2 + 18t - 4$

   (ii) $\cos t + \sin t$, $-\sin t + \cos t$

   (iii) $e^t \sin 2t + e^t 2\cos 2t$,
         $-3e^t \sin 2t + 4e^t \cos 2t$

   (iv) $\dfrac{dx}{dt} = \dfrac{2e^{2t}}{1+t} - \dfrac{e^{2t}}{(1+t)^2}$,

   $\dfrac{d^2y}{dt^2} = \dfrac{-2e^{2t}}{(1+t)^2} + \dfrac{2e^{2t}}{(1+t)^3}$

   (v) $\dfrac{dx}{dt} = \cos t\, e^{\sin t}$,

   $\dfrac{d^2x}{dt^2} = -\sin t\, e^{\sin t} + \cos^2 t\, e^{\sin t}$

8. (i) $v = 1$ m/s, $a = -4$ m/s²

   (ii) $v = 1$ m/s, $a = 1$ m/s²

   (iii) $v = 2$ m/s, $a = 4$ m/s²

   (iv) $v = 1$ m/s, $a = 2$ m/s²

   (v) $v = 1$ m/s, $a = 1$ m/s²

9. (i) $\dfrac{dy}{dx} = \dfrac{x^2 - 2x - 1}{(x-1)^2}$,

   $\dfrac{d^2y}{dx^2} = \dfrac{2}{x-1} - \dfrac{2(x^2 - 2x - 1)}{(x-1)^3}$

   (ii) $\dfrac{dy}{dx} = 2x\sin x + x^2 \cos x$,

   $\dfrac{d^2y}{dx^2} = 2\sin x + 4x\cos x - x^2 \sin x$

   (iii) $-\dfrac{\sin x + 3\cos x}{e^{3x}}$, $\dfrac{6\sin x + 8\cos x}{e^{3x}}$

   (iv) $x^{-1}(1+x)^{-2} - 2(1+x)^{-3}\ln x$,
        $-x^{-2}(1+x)^{-2} + x^{-1}\left[-2(1+x)^{-3}\right]$
        $+ 6(1+x)^{-4}\ln x - 2(1+x)^{-3}\dfrac{1}{x}$

   (v) $e^x(1+x)^{-1} - e^x(1+x)^{-2}$,

   $\dfrac{e^x}{1+x} - \dfrac{2e^x}{(1+x)^2} + \dfrac{2e^x}{(1+x)^3}$

10. (i) $\dfrac{d^2y}{dx^2} = 5$

    (ii) $\dfrac{d^2y}{dx^2} = 19.24$

    (iii) $\dfrac{d^2y}{dx^2} = 3.909$

    (iv) $\dfrac{d^2y}{dx^2} = 0$

11. $\dfrac{d^2y}{dx^2} = 75\,k^2$

12. See Text

13. $\dfrac{d^2y}{dx^2} = 4e^x \cos 2x - 3e^x \sin 2x$

15. $v = 216$ m/s, $a = 88$ m/s^2

16. (i) $\dfrac{dy}{d\theta} = \dfrac{1}{\cos\theta - 1}$,

    $\dfrac{d^2y}{d\theta^2} = \dfrac{\sin\theta}{(\cos\theta - 1)^2}$

    (ii) $\dfrac{dy}{dx} = \dfrac{\cos x - 2\sin x}{e^{2x}}$,

    $\dfrac{d^2y}{dx^2} = \dfrac{3\sin x - 4\cos x}{e^{2x}}$

    (iii) $\dfrac{dy}{d\theta} = -\mathrm{cosec}^2\theta$,

    $\dfrac{d^2y}{d\theta^2} = 2\,\mathrm{cosec}^2\theta \cot\theta$

17. $\dfrac{di}{dt} = 2\pi f I_m \cos\left(2\pi ft - \dfrac{\pi}{3}\right)$

    $\dfrac{d^2i}{dt^2} = -4\pi^2 f^2 I_m \sin\left(2\pi ft - \dfrac{\pi}{3}\right)$

18. $\dfrac{dV}{dt} = -2\pi f V_m \sin\left(2\pi ft + \dfrac{\pi}{6}\right)$

    $\dfrac{d^2V}{dt^2} = -4\pi^2 f^2 V_m \cos\left(2\pi ft + \dfrac{\pi}{6}\right)$

19. $\dfrac{dV}{dR} = 4\pi R^2$, $\dfrac{d^2V}{dR^2} = 8\pi R$

20. $\dfrac{dV}{dr} = 3\pi r^2$, $\dfrac{d^2V}{dr^2} = 6\pi r$

Exercises 8

1. $\dfrac{dy}{dx} = \cot\theta$, $y = x + 2a - \dfrac{a\pi}{2}$, $y = -x + \dfrac{a\pi}{2}$

2. $y = \left(\tan\dfrac{3t}{2}\right) x + 2\sin t - \sin 2t - 2\cos t \tan\dfrac{3t}{2}$
   $+ \cos 2t \tan\dfrac{3t}{2}$

   $y = -\left(\cot\dfrac{3t}{2}\right) x + 2\sin t - \sin 2t$
   $+ 2\cos t \cos\dfrac{3t}{2} - \cos 2t \cot\dfrac{3t}{2}$

3. $\dfrac{dy}{dx} = -\dfrac{3}{2}\cot\theta$,

   $y = \dfrac{2}{3}\tan\theta + \dfrac{5}{3}\sin\theta$

4. $y = \dfrac{1}{\sqrt{3}}x + \dfrac{5}{3}\sqrt{3}$, $y = -\sqrt{3}x + \dfrac{5}{2}\sqrt{3}$

5. $3y + 4 = x$, $3y + x = 7$

6. $y = x - 3$

7. $y = -9x - 18$, $9y + 80 = x$

Exercises 9

1. $\dfrac{dh}{dt} = \dfrac{2}{15} \times 10^{-3}$ m/s

2. $\dfrac{dr}{dt} = 3.98 \times 10^{-6}$ m/s, $\dfrac{dS}{dt} = 1 \times 10^{-3}$ m^2/s

3. $\dfrac{dr}{dt} = \dfrac{1}{8\pi 100}$, $\dfrac{dS}{dt} = 1$ mm^2/s, $\dfrac{dV}{dt} = 50$ mm^3/s

4. $\dfrac{dr}{dt} = 1.41 \times 10^{-3}$ mm/s

5. $\dfrac{dV}{dt} = 2.25$ m^3/s

6. $\dfrac{dy}{dt} = -0.060$ cm/s

7. $\dfrac{dV}{dt} = 6.28 \times 10^3$ cm^3/s

8. $\delta A = 0.866$ cm^2

9. (i) 1 (ii) k (iii) $\frac{2}{3}$ (iv) $\frac{1}{2}$

(v) 0 (vi) $-\infty$ (vii) $\frac{3}{4}$ (viii) $\frac{1}{2}$

(ix) $-\frac{3}{2}$ (x) e (xi) $\frac{3}{2}$ (xii) 2

(xiii) $-\frac{1}{2}$ (xiv) 0 (xv) ω

Exercises 10

1. (i) -0.820

   (ii) -2.71

2. 0.3517

3. 2.156

4. 1.433

Exercises 11

1. $\sin x = x - \frac{x^3}{3!} + \frac{x^5}{5!} - \cdots$

2. $\sin^{-1} x = x + \frac{1}{6}x^3 + \frac{3}{40}x^5$

3. $a_0 = 0, a_1 = 1, a_2 = 0, a_3 = -\frac{1}{3}, a_4 = 0,$ and $a_5 = \frac{1}{5}$

4. (a) $1 - x^2 + x^4 - x^6 + \cdots$

   (b) $1 - x^2 + x^4 - \cdots$

5. $\frac{2}{5}x + \frac{2}{375}x^3 + \frac{2}{15625}x^5 + \cdots$

6. $\sinh^{-1} \frac{x}{2} = \frac{1}{2}x - \frac{1}{48}x^3 + \frac{3}{1280}x^5 - \cdots$

7. $\sin^{-1} x = x + \frac{1}{6}x^3 + \frac{3}{40}x^5 + \cdots$

8. (i) $\sinh^2 x = x^2 + \frac{1}{3}x^4 + \frac{2}{45}x^6 + \cdots$

   (ii) $\log_e(1 - x^2) = -x^2 - \frac{x^4}{2} - \frac{x^6}{3} - \frac{x^8}{4} - \cdots$

   (iii) $\sin^2 x = x^2 - \frac{1}{3}x^4 + \frac{2}{45}x^6$

   (iv) $\cos^2 x = 1 - x^2 + \frac{1}{3}x^4 - \frac{2}{45}x^6$

9. (i) $\tan 45°2' = 1.0011642$

   (ii) $\log_e(1.001) = 9.985 \times 10^{-4}$

10. $\ln 1.0204 = 0.020195$

11. (i) $-4x - 8x^2 - \frac{64}{3}x^3 - 64x^4 - \cdots \quad \frac{-1}{4} \le x < \frac{1}{4}$

    (ii) $\ln 3 + \frac{5}{3}x - \left(\frac{5}{3}x\right)^2 \frac{1}{2} + \left(\frac{5}{3}x\right)^3 \frac{1}{3} - \cdots \quad \frac{-3}{5} < x \le \frac{3}{5}$

    (iii) $\ln 2 + \frac{7x}{2} - \left(\frac{7x}{2}\right)^2 \frac{1}{2} + \left(\frac{7x}{2}\right)^3 \frac{1}{3} - \left(\frac{7x}{2}\right)^4 \frac{1}{4} + \cdots \quad \frac{-2}{7} < x \le \frac{2}{7}$

12. $\pi = 4\left(1 - \frac{1}{3} + \frac{1}{5} - \frac{1}{7} + \cdots\right)$

13. $\pi = 6\left(\frac{1}{2} + \frac{1}{2 \times 3 \times 2^3} + \frac{3 \times 1}{2 \times 4 \times 5 \times 2^5} + \cdots\right)$

14. $\cos^{-1} x = \frac{\pi}{2} - x - \frac{1}{6}x^3 - \frac{3}{40}x^5 - \frac{5}{112}x^7 - \cdots,$
    $\pi = 3.141$ to 3 d.p.

4. DIFFERENTIAL CALCULUS AND APPLICATIONS

Index

A
Algebraic functions 1
Angle between two lines 40
Approximations rates 48
Approximate solutions of equations 57

C
Circle 34
Circular functions 13
Concept of a limit 1

D
Derivatives
  of a difference of function 4
  of an implicit function 9
  of a logarithms function 24
  of a product of a function 5
  of a quotient of a function 7
  of a sum function 4
  of a function of a functions 8
  of tangent 13
  of cotangent 13
  of cosecant 14
  of secant 14
Differentiation 1
Differential calculus 1
Differentiation from first principles
  algebraic functions 2
  exponential functions 21
  trigonometric functions 13

E
Ellipse 34
Equations of normal and tangents 40
Exponential functions 20

F
First Derivative 2
Function of a function 8, 28
Function of a function of a function 17
Functions Algebraic 1

G
Gradient (Notation) 2

H
Higher derivatives of a function hyperbola 34
Hyperbolic
  functions 27
  graphs 30–1

I
Implicit functions 17
Improvement of an approximation 51–3
Inverse
  hyperbolic function 29
  graphs 30–1
  trigonometric functions 17

L
Leibnitz's theorem 63
L'Hôpital's rule 45
Limit (the concept) 1
Logarithmic functions 24-5

M
Maclaurin's expansion 55

N
Newton-Raphson's method 51–3
Normals 40
Notation of a gradient 2
Numerical methods for the solution of differential
    equations 63

P
Parabola 34
Parametric equations 33
Polynomial approximations using Taylor's series 57
Power series 55
Primitive 2, 37–8

Q
Quotient Rule 7

R
Rates of change 37, 48
Rectangular hyperbola 34

S
Second derivatives 37
Slope 2
Small increments and approximations 45
Stirling's 60
Successive approximation 60

T
Tangents & Normals 40–4
Taylor's series 57
Trigonometric functions 13
Turning points $\frac{dy}{dx} = 0$ 37–8